Copyright, 1910, by The Pictorial News Co.
CURTISS' HUDSON RIVER FLIGHT—OVER THE STATUE OF LIBERTY

THE CURTISS AVIATION BOOK

BY

GLENN H. CURTISS

AND

AUGUSTUS POST

WITH CHAPTERS BY CAPTAIN PAUL W. BECK, U. S. A.
LIEUTENANT THEODORE G. ELLYSON, U. S. N.
AND HUGH ROBINSON

With Numerous Illustrations from Photographs

NEW YORK
FREDERICK A. STOKES COMPANY
PUBLISHERS

TO
MRS. MABEL G. BELL
WHO MADE POSSIBLE THE AERIAL EXPERIMENT ASSOCIATION
THIS BOOK IS DEDICATED BY
THE AUTHORS

CONTENTS

PART I

BOYHOOD AND EARLY EXPERIMENTS—
Augustus Post

CHAPTER		PAGE
I	THE COMING AIRMEN—AN INTRODUCTORY CHAPTER	3
II	BOYHOOD DAYS	8
III	BUILDING MOTORS AND MOTORCYCLE RACING	18
IV	BALDWIN'S BALLOON	29

PART II

MY FIRST FLIGHTS—*Glenn H. Curtiss*

I	BEGINNING TO FLY	37
II	FIRST FLIGHTS	41
III	THE "JUNE BUG"—FIRST FLIGHTS FOR THE SCIENTIFIC AMERICAN TROPHY AND FIRST EXPERIMENTS WITH THE HYDROAEROPLANE	51
IV	FIRST FLIGHTS IN NEW YORK CITY	57

PART III

MY CHIEF FLIGHTS AND THE WORK OF TO-DAY—*Glenn H. Curtiss*

I	THE RHEIMS MEET—FIRST INTERNATIONAL AEROPLANE CONTEST	65

CONTENTS

CHAPTER		PAGE
II	HUDSON-FULTON CELEBRATION—FIRST AMERICAN INTERNATIONAL MEET, LOS ANGELES	80
III	FLIGHT DOWN THE HUDSON RIVER FROM ALBANY TO NEW YORK CITY	91
IV	THE BEGINNING OF THE HYDROAEROPLANE	112
V	DEVELOPING THE HYDROAEROPLANE AT SAN DIEGO—THE HYDRO OF THE SUMMER OF 1912	129

PART IV

THE REAL FUTURE OF THE AEROPLANE—
Glenn H. Curtiss, Capt. Beck, Lieut. Ellyson and Augustus Post

I	AEROPLANE SPEED OF THE FUTURE	155
II	FUTURE SURPRISES OF THE AEROPLANE—HUNTING, TRAVEL, MAIL, WIRELESS, LIFE-SAVING, AND OTHER SPECIAL USES	168
III	THE FUTURE OF THE HYDRO	187
IV	FUTURE PROBLEMS OF AVIATION	193
V	THE AEROPLANE AS APPLIED TO THE ARMY—*Capt. Paul W. Beck, U. S. A.*	205
VI	THE AEROPLANE FOR THE NAVY—*Lieut. Theodore G. Ellyson, U. S. N.*	219
VII	GLIDING AND CYCLE-SAILING—A FUTURE SPORT FOR BOYS, THE AIRMEN OF TO-MORROW—*Augustus Post*	227

PART V

EVERY-DAY FLYING FOR PROFESSIONAL AND AMATEUR—*Glenn H. Curtiss, Augustus Post and Hugh Robinson*

I	TEACHING AVIATORS—HOW AN AVIATOR FLIES	235
II	AVIATION FOR AMATEURS	254

CONTENTS

CHAPTER		PAGE
III	How it Feels to Fly—*Augustus Post*	263
IV	Operating a Hydroaeroplane—*Hugh Robinson*	272

PART VI

THE CURTISS PUPILS AND A DESCRIPTION OF THE CURTISS AEROPLANE AND MOTOR—*Augustus Post*

I	Pupils	281
II	A Description of the Curtiss Biplane	287
III	The Curtiss Motor and Factory	296

ILLUSTRATIONS

CURTISS' HUDSON RIVER FLIGHT—OVER THE STATUE OF LIBERTY . . . . *Frontispiece*	
	FACING PAGE
CURTISS THE BOY AND CURTISS THE MAN	20
CURTISS WINNING WORLD'S MOTORCYCLE RECORDS . . .	21
THE BALDWIN ARMY DIRIGIBLE, WITH EARLY CURTISS MOTOR	30
WIND WAGON AND ICE BOAT WITH AERIAL PROPELLER . .	31
THE AERIAL EXPERIMENT ASSOCIATION	38
STARTING TO FLY—FIRST PUBLIC FLIGHT IN AMERICA; THE "JUNE BUG," JUNE, 1908; BALDWIN IN GLIDER . . .	39
THE FIRST MACHINES—THE "WHITE WING" AND "RED WING"	52
CURTISS' FIRST FLIGHT FOR THE SCIENTIFIC AMERICAN TROPHY	53
WINNING THE GORDON BENNETT CONTEST IN FRANCE . .	74
PRESIDENT TAFT WATCHING CURTISS FLY, HARVARD MEET, 1910	75
THE ALBANY-NEW YORK FLIGHT—START; OVER WEST POINT	92
THE HUDSON FLIGHT—OVER STORM KING	93
THE HUDSON FLIGHT—STOP AT POUGHKEEPSIE; FINISH, AT GOVERNOR'S ISLAND	106
THE EVOLUTION OF THE HYDRO;—THE FIRST HYDRO IN THE WORLD; DUAL CONTROL HYDRO OF 1911; LANDING IN HYDRO AT CEDAR POINT, OHIO	107
ELY LANDING ON THE U. S. S. "PENNSYLVANIA"	120
CURTISS AND HYDRO HOISTED ON U. S. S. "PENNSYLVANIA"; ELY LEAVING "PENNSYLVANIA"	121

ILLUSTRATIONS

	FACING PAGE
DIAGRAM OF CURTISS FLYING BOAT OF 1912	146
THE EVOLUTION OF THE HYDRO—THE FLYING BOAT OF SUMMER 1912; THE 1911 HYDRO	148
HYDRO FLIGHTS—CURTISS OVER LAKE ERIE; WITMER RIDING THE GROUND SWELLS	149
CAPTAIN BECK AND POSTMASTER-GENERAL HITCHCOCK CARRYING THE MAIL	174
STUDENTS OF AERIAL WARFARE—BECK, TOWERS, ELLYSON, McCLASKEY; WITH CURTISS AND ST. HENRY	175
ELLYSON LAUNCHES HYDRO FROM WIRE CABLE	224
HUGH ROBINSON'S FLIGHT DOWN THE MISSISSIPPI	225
AUGUSTUS POST FLYING; AEROPLANE SHIPMENT	264
CURTISS PUPILS—J. A. D. McCURDY RACING AN AUTOMOBILE; LIEUTENANT ELLYSON; MR. AND MRS. W. B. ATWATER	265
CURTISS PUPILS—C. C. WITMER, BECKWITH HAVENS, J. A. D. McCURDY, CROMWELL DIXON, CHAS. K. HAMILTON, CHAS. F. WALSH, CHAS. F. WILLARD	282
LINCOLN BEACHEY FLYING IN GORGE AT NIAGARA	283
DIAGRAM OF CURTISS AEROPLANE, SHOWING PARTS	290
DIAGRAM OF CURTISS MOTOR, SHOWING PARTS	291
CURTISS MOTORS, OLD AND NEW	300
AT THE AEROPLANE FACTORY, HAMMONDSPORT	301

PART I

BOYHOOD AND EARLY EXPERIMENTS OF GLENN H. CURTISS

BY
AUGUSTUS POST

THE CURTISS AVIATION BOOK

CHAPTER I

THE COMING AIRMEN—AN INTRODUCTORY CHAPTER

THE time has come when the world is going to need a new type of men—almost a new race. These are the Flying Men. The great dream of centuries has come true, and man now has the key to the sky. Every great invention which affects the habits and customs of a people brings about changes in the people themselves. How great, then, must be the changes to be brought about by the flying machine, and how strangely new the type of man that it carries up into a new world, under absolutely new conditions!

Each year there will be more need of flying men; so that in telling this story of a pioneer American aviator, his struggles, failures, and successes, it has been the desire to keep in mind not only the scientific elders who are interested in angles of incidence, automatic stability and the like, but also the boys and girls—the air pilots of the future. It is hoped that there will be in these introductory chapters—for whose writing, be it understood, Mr. Curtiss is not responsible—a plain unvarnished story of an American boy who worked his way upward from the making of bicycles to the making of history, an inspiration for future flights, whether in imagination or aeroplanes, and that even the youngest reader will gain courage to meet the obstacles and to overcome the difficulties which Glenn H. Curtiss met and overcame in his progress to fame.

Here is a man who is a speed marvel—who has beat the world at it. First on land, riding a motorcycle, next in a flying machine, and finally in a machine that was both water and air craft, which sped over the surface of the sea faster than man had ever travelled on that element, and which rose into the air and came back to land with the speed of the fastest express train; a man who traveled at the rate of one hundred and thirty-seven miles an hour on land, fifty-eight miles an hour on the water and who won the first International speed championship in the air.

More than that, they may see what sort of a boy came to be the speed champion and to know some of the traits that go to make the successful airman, for it is said of the great aviators, as of the great poets, they are born flying men, and not developed. The successful flying man and maker of flying machines, such as Glenn H. Curtiss has shown himself to be, realises how dangerous is failure, and builds slowly. He builds, too, on his experience gained from day to day; having infinite patience and dogged perseverance. And yet a great aviator must be possessed of such marvelous quickness of thought that he can think faster than the forces of nature can act, and he must act as fast as he thinks.

He must be so completely in harmony with Nature and her moods that he can tell just when is the right time to attempt a dangerous experiment, and so thoroughly in control of himself that he can refuse to make the experiment when he knows it should not be made, even though urged by all those around him to go ahead. He must feel that nothing is impossible, and yet he must not attempt anything until he is sure that he is ready and every element of danger has been eliminated, so far as lies in human power. He must realise that he cannot change the forces of nature, but that he can make them do his work when he understands them. Some of these qualities must be inbred in the man, but the life-story of Glenn H. Curtiss shows how far energy, courage, and tireless perseverance will go toward bringing them out.

It is from among the country boys that the best aviators will be found to meet the demands of the coming Flying Age. They have been getting ready for it for a long time—long before the days of Darius Green. Does any one now read "Phaeton Rogers,'' that story of the inventive boy back in the eighties, and recall the "wind-wagon" which was one of his many inventions? There were many like him then, and there are more like him now; always tinkering at something, trying to make it "go," and go fast. And there are many of these who are building up, perhaps without knowing it, the strong body, the steady brain, courage, perseverance, and the power of quick decision—the character of the successful airmen of the future.

The history of aviation is very brief, expressed in years. In effort it covers centuries. First come the inventors, a calm, cautious type of men, holding their ideas so well in trust that they will not risk their lives for mere display and the applause of the crowd. Then the exploiters, eager for money and fame; men who develop the possibilities of the machines, always asking more and getting more in the way of achievement with each new model built. Though covering a period of less than a half score of years, aviation already has its second generation of flyers, pupils trained by the pioneers, young and ambitious, eager to explore the new element that has been made possible by their mentors. From the country districts, where the blood is red, the brain steady and the heart strong, will come many an explorer of the regions of the air. Just as the city boy in developing the wireless telegraph strings his antennæ on the housetops and the roofs of the giant skyscrapers, so will the country boy develop his glider or his aeroplane in the pasture lands and on the steep hillsides of his own particular territory, and we shall have a race of flying men to carry on the development of the flying machine until it shall reach that long dreamed-of and fought-for perfection.

CHAPTER II

BOYHOOD DAYS

GLENN HAMMOND CURTISS was born at Hammondsport, New York, May 21, 1878. His middle name shows his connection with the pioneer family for which the town is named. Then Hammondsport was a port for canal boats that came up Lake Keuka; nowadays it is an airport for the craft of the sky. It is a quaint little town, lying on the shores of a beautiful lake that stretches away to Penn Yann, twenty miles to the north. Glenn's old home was called Castle Hill. It was nearly surrounded by vineyards and fruit trees. It was once the property of Judge Hammond, who built the first house in Hammondsport. On this site now stands the Curtiss factories.

All about Hammondsport are the great vineyards that have made the town famous for its wine, for Hammondsport is in the very heart of the grape-growing section of New York State. These vineyards give the boys of Hammondsport a fine opportunity to earn money each year, and Glenn was always among those who spent the vacation time in tying up grape vines, and in gathering the fruit on Saturdays and at other odd times.

Some of the neighbours' children picked wintergreen and flowers, and sold them to the summer excursionists. One time Glenn was invited to go with them. He sold six bunches for sixty cents. His mother applied the amount toward a pair of shoes in order to teach him the use and value of money. He was then three years old and wore a fresh white dress and a blue sash.

Glenn was afterwards taught how to prune and tie vines and gather fruit and at harvest time he was often seen with pony and wagon making a fast run to the station to get the last load of grapes on the train.

With the care of his sister and the work on the home vineyard, life was not all play, for Glenn was "The Man of the House," after his father's death, which occurred when he was four years old. At this time, he went with his mother and sister, to live with his grandmother who lived on the outskirts of the village.

Hammondsport is divided by the main street, and the boys of the two sections, like the boys in cities, were always at war. The factional lines were tightly drawn and many were the combats between the up-town boys and the low-town boys. The hill boys had a den in the side of a bank that sloped down from Grandma Curtiss' yard, walled in with stones of a convenient size. This gave them good ammunition and a great advantage in time of battle.

Among the members of the up-town gang were, "Fatty" Hastings and "Short" Wheeler, "Jess" Talmadge and "Cowboy" Wixom and Curtley, as the boys called Curtiss. He was captain of the band, because he had a sort of ownership of the den. Thus the war waged until one day they punctured Craton Wheeler's dog "Pickles," which so infuriated the enemy of the lower village that they were on the point of storming the fort in the hillside from above, and would no doubt have done so had they not chanced to trample upon Grandma Curtiss' flower beds which caused this indignant lady to issue forth and put the entire gang to rout. The cave continued to be a safe refuge for the hillside gang until "Fatty" Hastings grew too big to squeeze through the entrance and sometimes got stuck just as the gang was ready to sally forth against the enemy, or blocked the whole crew when they were in retreat.

During the winter months Glenn gave his hand to making skate-sails, and became very proficient at it, and when summer came and the boys went on bird-nesting excursions in the woods, he was usually the daring one who allowed himself to be lowered by a rope over the cliff's edge or climbed to the topmost limbs of the big hickory trees. At school, mathematics was young Curtiss's strong point, and when finally he came to pass his final examinations in the high school, he topped his class in that study with a perfect score of one

BOYHOOD DAYS

hundred, and in Algebra he stood ninety-nine. It is reassuring, however, to find that in spelling he was barely able to squeeze through with a percentage of seventy-five. Glenn sometimes slipped up on the figuring, but the principle was usually right; he had figured that out beforehand. The boys of Hammondsport used to say that Glenn would think half an hour to do fifteen minutes' work. One wonders what they would have said, if they had been told that in after years he was to think and plan and scheme for a year, and then when he was all ready, to wait hour after hour, day after day, to accomplish something requiring a little more than two hours' time; like his flight from Albany to New York, the first great cross-country flight made in America.

When Curtiss was twelve years old his family went to live in Rochester, New York, so that his sister might be able to attend a school for the deaf at that place. He went on working at Rochester after school hours and during vacation time, first as a telegraph messenger, then in the great Eastman Kodak works, assembling cameras. He was one of the very first boys hired by that establishment to replace men at certain kinds of work, and while the men had received twelve dollars a week, Glenn received but four dollars. Before long, however, he had induced his employers to make his work a piece-work job, and had improved the process of manufacture and increased the production from two hundred and fifty to twenty-five hundred a day. He was thus able to earn from twelve to fifteen dollars a week. It was while employed in the camera works at Rochester that Curtiss saved the life of a companion who had fallen through the ice on the Erie canal. When praised for his act of bravery he simply remarked: "I pulled him out because I was the nearest to him."

All during the time that Curtiss was working for others for wages, he continued to tinker—making things and then taking them apart. Once he told some of his companions that he could make, out of a cigar box, a camera that would take a good picture. Of course they laughed at him and bet that he couldn't do it. But Glenn did do it, and a picture of his sister with a book was produced and is still unfaded, and in good condition, in possession of his family. He constructed a complete telegraph instrument out of spools, nails, tin, and wire and this so impressed the lady with whom the Curtisses boarded that she remarked to one of her friends that "Glenn Curtiss will make his mark in the world some day; you mark my words." This particular lady tells of the time that Glenn used to talk of airships, and he was not yet sixteen years old. Curtiss was fond of all sorts of sports, taking part in the games the boys would get up after school and on Saturdays. He

BOYHOOD DAYS

liked to play ball, to run, jump, swim, and to ride a bicycle.

His time was too much taken up, however, with more productive efforts, such as the wiring of dwellings for electric light or telephones, to permit of much time being given to boyish sports.

He was most original and had a keen sense of humour. He was fond of an argument, and had one striking characteristic; once he had made up his mind as to the why and wherefore of a thing, he could never be induced to change it. To illustrate this trait; one day an argument arose between Glenn and another boy as to whether or not a whale is a fish, Glenn holding that it could be nothing but a fish. The other boy finally reënforced his argument by producing a dictionary to show that a whale is not a fish, whereupon Curtiss asserted that the dictionary was wrong and refused to accept it as authority.

Curtiss was always eager for speed—to get from one place to another in the quickest time with the least amount of effort. He was obsessed with the idea of travelling fast. One of the first things he remembers, says Curtiss, was seeing a sled made by one of his father's workmen for his son beat every other sled that dashed down the steep snow-clad hills around Hammondsport. He begged his father to let "Gene" make him a sled that would go faster than Linn's. "Gene" made the sled and Glenn painted it red, with a picture of a horse on it. Furthermore, he beat every sled in Hammondsport or thereabouts.

The bicycle became all the rage when Curtiss was growing into his early teens and nothing was more certain than that he should have one as soon as he could earn enough money to buy it. And when he got it he made it serve his purposes in delivering telegrams, newspapers, and such like. He developed speed and staying powers as a rider, and soon thought nothing of making the trip from Rochester to Hammondsport to see his grandmother, who still lived in the old home in that village. The roads of New York were not as good as they are nowadays, when the automobile forces improvements of the highways, but Curtiss rode fast nevertheless. In fact, he managed all his regular work this way. His idea was first, to find out just how to do it, and then do it. Then he would find out how fast a certain task could be performed, and get through with it at top speed. The surplus time he devoted to tinkering with something new.

Grandmother Curtiss finally prevailed upon him to go back to Hammondsport and live with her. For a time after his return he assisted a local photographer and his experience in photography gained at this time has since proved of great value to him, and, incidentally, to the history of aviation; for in photographing his experiments

Curtiss' pictures have a distinct value, as much for being taken just at the right instant, as for their pictorial detail. Following his photographic employment, Curtiss took charge of a bicycle repair shop. It was a little shop down by the principal hotel in Hammondsport, but Curtiss foresaw the popularity and later the cheapness of the bicycle, and he believed the shop would do a good business. James Smellie owned the shop, but Curtiss' mechanical skill soon asserted itself and he became the practical boss. This was in 1897. George Lyon, a local jeweler, was a competitor of Smellie's in the bicycle business, and got up a big race around the valley, a distance of five miles over the rough country roads. When Smellie heard of the race he made up his mind that Curtiss could win it and went about arranging the equipment of his employé. That race has passed into the real history of the town of Hammondsport. Everybody in the town and the valley was there, and great was the excitement when the riders lined up for the start. They started from a point near the monument in front of the Episcopal church and within a few moments after the crack of the pistol they were all out of sight, swallowed up in the dust clouds that marked their progress up the valley. After a long interval of suspense a solitary rider appeared on the home stretch, hunched down over his handle-bars and riding for dear life, without a glance to right or left. It was Curtiss, who probably has never since felt the same thrill of pride at the shouts of the crowd. The next man was fully half a mile in the rear when Glenn crossed the finish-line.

This was Curtiss' first bicycle race, but later he acquired greater speed and experience and rode in many races at county fairs in the southern part of New York State. What's more, he won all of his races. This was good for his bicycle business, which thrived in the summer, but languished in the winter. During the dull period Curtiss took up electrical work, wiring houses, putting in electric bells, and doing similar work of a mechanical nature. An incident is told of his mechanical skill at this time that illustrates his inquisitive mind. An acetylene gas generator in one of the stores got out of order one day, and no one in the store could tell just how to repair it. Curtiss had never seen a gas generator, but that did not deter him from going at it. He studied it out in a little while and then put his finger on the trouble. After that the generator worked better than ever. A little later he decided to build a gas generator after his own ideas. He started with two tomato cans and built it.

This was the first appearance of Curtiss' two tomato cans. They played an important part in his subsequent experimental work, figuring all the way through from this first gas generator to the carburetor of a motorcycle, and at last to enlarge

BOYHOOD DAYS

the water capacity of Charles K. Hamilton's engine on his aeroplane so that he might cool his engine better in making the record flight from New York to Philadelphia and return in the same day. In this first case the two tomato cans developed into an acetylene gas plant with several improvements, and his own home and shop were lighted by it. Later the plant was enlarged so as to furnish light for several business houses of Hammondsport.

CHAPTER III

BUILDING MOTORS AND MOTORCYCLE RACING

IN the spring of 1900 Curtiss embarked in the bicycle business for himself, opening a shop near his old place of employment. This shop soon came to be known as the "industrial incubator," because experiments of many kinds were tried there—a hatching-place for all sorts of new machines. The first one developed was destined to open up to Curtiss a new field of action, one that furnished the opportunity for new speed records, and enlarged the scope of his activities beyond the limits of the little town and the valley, and spread before him possibilities as wide as the boundaries of the continent.

Curtiss had ridden a bicycle in races, and got the utmost speed out of it; but the bicycle, as a man-propelled vehicle, did not travel fast enough to suit him. He therefore set about devising means for increasing its speed possibilities. One day Smellie, his old employer, came into Curtiss' shop, tired out and perspiring from his efforts in pedaling his bicycle up the hill. "Glenn," he said, "I'm going to give the blamed thing up until they get something to push it." That was Cur-

MOTORS AND MOTORCYCLES 19

tiss' cue, and it promptly became his problem—getting something to push it! He determined to mount a gasoline engine on a bicycle, and at once began to search for the necessary castings. Finally he secured them and began the task of building a motor. Unfortunately, the man who sold him the castings sent no instructions for building a motor, so the problem was left to Curtiss and to those who interested themselves in his work. They studied and planned and made experiments, learning something new about motors all the while. Eventually, with the assistance of local mechanics, the castings were "machined" and the motor assembled.

Curtiss afterward described it as a remarkable contrivance; but it did the work. This motor had a two-inch bore and a two-an-a-half-inch stroke, and drove the bicycle wheel by a friction roller pulley. First, Curtiss made the pulley of wood, then of leather, and finally of rubber. It was tried first on the front wheel and then on the rear one, and so numerous were the changes in and additions to its equipment, that the bystanders—and there was the usual number of these—saw only the humorous side of the thing and declared that it looked like a sort of Happy Hooligan bicycle with tin cans hung on wherever there was room. The tomato can again came to the front in Curtiss' experiments, and now served to fashion a rough and ready sort of carburetor, filled with gasoline and covered over with a gauze screen, which sucked up the liquid by capillary attraction. Thus it vaporized and was conducted to the cylinder by a pipe from the top of the can.

Then came the first demonstration of a bicycle driven by power other than leg muscles, and it attracted almost as much attention in Hammondsport as the first bicycle road race which Curtiss had won some years before. The newfangled machine, which the village oracle declared could not be made to go unless the rider put his legs to work, did not promise much of a success on its initial trip. Curtiss started off for the post-office, but had to pedal all the way there, the motor refusing to do its part. Coming from the post-office, however, it began popping and shoved the wheels around at an amazing rate, while Curtiss sat calmly upright and viewed the excited citizens of Hammondsport as he sped by.

That was the beginning of Curtiss' motorcycle; but the ambitious inventor did not rest with the first success. Work at the "incubator" went on unceasingly. The young mechanical genius carried on his regular duties during the days but spent most of the nights in his experiments. Curtiss would not have said that he worked nights, but that he spent his evenings in "doping out" the best way to build something. He has never changed his habits in this respect. He still "dopes out" something for the next day or the

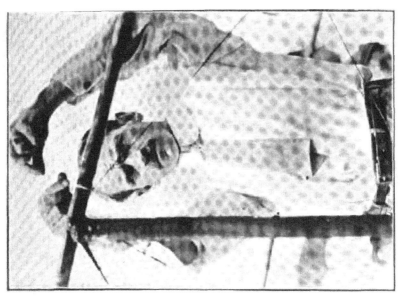

(A) POST CARD SENT BY CURTISS TO HIS WIFE, JANUARY 24, 1907
(B) CURTISS MAKING WORLD'S MOTORCYCLE RECORD, ORMOND BEACH

next month while "resting" from his daylight duties; though the process would now be expressed in somewhat more scientific terms. In truth, one may say that Curtiss worked all the time. In office or shop hours, like other persons, he did what he *had* to do; while at other times he did what he *wanted* to do. Curtiss was different only in that he *wanted* to do those things which other people would call labor. Experimental work was recreation to Curtiss, and because of this mental attitude he was able to stick at a task day and night and keep up "steam" all the while.

Curtiss seldom planned on paper. Plans seemed to outline themselves in his active mind, and when, later, he became an employer of a number of men, he simply outlined his ideas, describing just what he wanted to accomplish, and left it to their ingenuity. Sometimes one of his assistants would ask him a question and after standing for minutes as if he had not heard, Curtiss would suddenly reply and outline a task which it would require all day to carry out. Once Curtiss had decided that a certain course of action would bring certain mechanical results, it usually turned out that way, and because of this and the further fact that he was as good a workman as he was a designer, the men he had gathered around him grew to regard his judgment as final and therefore went ahead with absolute confidence as to the results.

There was a remarkable spirit of coöperation in the "industrial incubator." This spirit continued through the early years of Curtiss' first business successes, and it obtains to-day in the big Curtiss aeroplane and motor factories at Hammondsport. The alertness of the men around Curtiss, and the atmosphere of coöperation may be due, in some measure, to the curious interest they always hold as to what he will do next—and there is certain to be something happening out of the ordinary. Thus, work with Curtiss seldom becomes monotonous and without its surprises.

To go back to the first motor Curtiss built; it was quickly found to be too small, and he secured another set of castings, as large as he could get. With these he constructed a motor with a cylinder three and a half by five inches, and weighing a hundred and ninety pounds. This machine proved to be.a terror. It is true that it exploded only occasionally, but when it did it almost tore itself loose from the frame. But it drove the motorcycle as fast as thirty miles an hour and gained such a remarkable reputation in Hammondsport that a story is still told in the town of the time Curtiss made his first trip with it, when it carried him through the village, up over the steep hills, through North Urbana and as far as Wayne, where it ran out of gasoline and came to a stop of its own accord.

Thus Curtiss went ahead with his work to con-

MOTORS AND MOTORCYCLES

struct and improve his motors, and improvement came with each successive one. The third motor was better suited to the needs of the bicycle and furnished better results. Meantime, Curtiss began to receive inquiries and even some orders, and business took a decidedly favorable turn. Judge Monroe Wheeler took a great liking to the young man, who used to come over to his office to get the judge's stenographer to typewrite his letters, and helped him to establish credit at the local bank, and in other ways. Half a dozen fellow-townsmen became interested enough in Curtiss' motorcycle experiments to put money into the business, and within a short time a little factory was built on the hill back of Grandma Curtiss' house. It was an inconvenient place to put up a factory, and all the heavy material was hauled up to it with some difficulty, but the light, finished product, which in this case could go under its own power, rolled down the steep grade without trouble. In spite of these little obstacles; in spite of the fact that Hammondsport is located at the end of a little branch railroad which seems to the visitor to run only as the spirit moves the engineer—in spite of every handicap, the business grew rapidly.

Curtiss was, by this time, happily married and Mrs. Curtiss helped with the office work at the factory, which stood then, as it does to-day, at the very back door of the old Curtiss homestead on the hillside. Curtiss used to take out his best motorcycle in these days and go off alone to all the motorcycle races held in that section of the State. Incidentally, he scooped in all the prizes, for he had the fastest machine, and he was a finished rider. On Memorial Day in 1903, Curtiss ventured far afield for an event that brought him his first notices in the big newspapers of New York City. He entered and won a hill-climbing contest at New York City, on Riverside Drive, and immediately afterward mounted his wheel, rode up the Hudson to another race, at Empire City Track, and won that also. This gave him the American championship.

Later, at Providence, R. I., he established a world's record for a single-cylinder motorcycle, covering a mile in fifty-six and two-fifths seconds. While this was phenomenal speed, it was as nothing in comparison with the record he was soon to establish. He built a two-cylinder motor and on January 28, 1904, at Ormond Beach, Florida, he rode ten miles in eight minutes fifty-four and two-fifth seconds, and established a world's record that stood for more than seven years. Curtiss was not content even with this. He wanted to travel faster than man had ever traveled before. He had built a forty horse-power, eight-cylinder motor for a customer who wanted it to put in a flying machine which he was building, and in order to try out the motor Curtiss built an especially strong motorcycle, using an automobile tire on the rear wheel and a motorcycle tire on the front wheel. On a strong frame the big forty horsepower motor was mounted. It was not given a thorough try out at Hammondsport, for it was winter and snow lay deep on the roads. With the aid of some of his shopmen, Curtiss took the freak machine out on the snow-covered roads, merely for the purpose of seeing if it could be started as it was geared in the machine. It proved that it would start all right, and so it was hurriedly boxed and rushed to the train, which was actually kept waiting several minutes. Curtiss was going South to make new records, and even the railroad men on the little branch road from Hammondsport to Bath, felt an interest in his undertaking. This, by the way, is typical of the way things are done at Hammondsport. When there is need for rushing matters, the men work night and day without complaint. These last-moment rushes are often due to the giving of much thought to the details before commencing to build, and sometimes because, in building, improvements which must be incorporated suggest themselves. Curtiss' rule, as he expresses it, is: "What is the need of racing unless you think you are going to win; and if you are beaten before you start, why take a chance?" But there are other considerations for the builder of racing machines to take into account. If your competitors know what you are doing, and they will know, somehow, if you give them a little time, they will go you one better. Therefore, this belated activity at the Curtiss factory is not always without its motive. Take, for instance, the first big International race for the Gordon Bennett aviation trophy, which Curtiss won at Rheims, France, in 1909. In spite of the fact that Curtiss' motor was built in a great hurry, barely giving the necessary time to finish it and reach Rheims for the race, Blériot, the chief French builder of the monoplane type, changed his motor as soon as he had read a description of the one Curtiss was to use.

The motorcycle which Curtiss had built and mounted with the eight-cylinder motor proved to be a world-beater—the fastest vehicle ever built to carry a man. It was taken to Ormond Beach, Florida, where it was tried out on the smooth sandy shore, which stretches for miles, as level as a billiard table and almost as hard as asphalt. Here, on January 24, 1907, Curtiss mounted the heavy, ungainly vehicle and traveled a mile in twenty-six and two-fifth seconds, at the rate of one hundred and thirty-seven miles an hour! This stands to-day as the speed record for man and machine. Curtiss, without goggles and with no special precautions in the matter of costume, simply mounted the seat, took a two-mile running start before crossing the line, and was off. Bending so low over the handle-bars that he almost

MOTORS AND MOTORCYCLES 27

seemed to be lying flat and merged into a part of the machine itself, he flashed over the mile course in less time than it takes to read these dozen lines. This speed trial was the culmination of weeks of study, work, and experiment. Day after day, and even at night, Curtiss had schemed and worked; now to get the weight properly placed and balanced; here to strengthen the frame and overcome the danger from the torque, and the tendency to turn the machine over, and finally to obtain the right sort of tires and to put them on securely. Ordinary tires, on wheels revolving at such an amazing speed, would have been cast off the rims like a belt off a pulley, by the centrifugal force.

These and a thousand other details were worked out so thoroughly that the machine, when ready, required very little testing out. In describing the trial Curtiss said that he could see nothing but a streak of grey beach in front of him, a blur of hills on one side, and the white ribbon of foaming surf on the other. The great crowd that watched the smoking, whirring thing that flashed by as if fired from a great gun, caught but a fleeting glimpse of Curtiss.

The record could not be accepted as official, because the motor was too big and powerful to be classed as a motorcycle engine. It therefore stands as an absolutely unique performance, unequalled, and not even approached as regards speed, until three years later, when Barney Oldfield, driving a two hundred horse-power Benz automobile, covered a mile over the same course in twenty-seven and thirty-three hundredths seconds.

Curtiss had developed, improved, and exhausted the motorcycle as far as speed possibilities were concerned, and was soon to give it up for something of far greater potential possibilities—the aeroplane.

CHAPTER IV

BALDWIN'S BALLOON

THOMAS SCOTT BALDWIN was engaged in building a dirigible balloon in California when he chanced to see a new motorcycle, the motor of which seemed to be exactly what he wanted to propel his new airship. He learned that it was the design and product of a man named Curtiss, at Hammondsport, N. Y., with whom he entered into correspondence. The result was that Captain Baldwin went to Hammondsport for a personal interview with the man who had turned out the motor.

Baldwin expected to find, as he afterward said, a big, important-looking manufacturer, and great was his surprise to find a quiet, unassuming young man, scarcely more than a youth. The jovial Baldwin and the unobtrusive Curtiss became great friends at once. They discussed motors of all sorts, but particularly motors suitable for dirigible balloons, then in the first stage of development. When Baldwin asked Curtiss the price of one of the type then used in the Curtiss motorcycle, he was surprised at its cheapness, and ordered one on the spot. This was built at once and proved successful. Later several other motors were built at the Curtiss factory for Baldwin, each one showing some improvement, and some of them designed to meet the increasing demand for a more powerful motor of light weight for use in dirigible balloons. As a natural consequence of Baldwin's success with the use of the Curtiss motor, it was but a short time until it came to be the best known motor in America for aeronautic work. At the St. Louis World's Fair, in 1904, Captain Baldwin's "California Arrow," the only successful airship out of all those which were brought from Europe and every part of America to contest for big prizes, was equipped with one of Curtiss' motors. Baldwin's success at St. Louis was a triumph for Curtiss, and soon all dirigible balloons operating in this country were driven by Curtiss motors.

Hammondsport was now to have a new sensation and to witness an experiment which eventually led to momentous developments. In order to test the power of the motors he was building for Captain Baldwin, and for the purpose of determining the efficiency of his aerial propeller, Curtiss constructed a "wind-wagon," a three-wheel vehicle with the motor and propeller mounted in the rear of the driver. When he took this queer contrivance out on the road for its first trial, the town of Hammondsport turned out to witness the fun. Consternation among the usually mild-eyed work

NEARLY UP IN THE AIR
(A) The wind wagon—Curtiss in 1904. (B) Ice boat with aerial propeller

horses spread throughout the little valley as the "wind-wagon" went scooting up and down the dusty roads, creating a fearful racket. Before the start was made an automobile was sent ahead to clear the way and to warn the drivers of other vehicles. The automobile, however, was quickly overhauled, passed, and left far in the rear by the whirring, spluttering, three-wheeled embryonic flying machine.

Protests by farmers, business-men and others quickly followed this experiment. They argued that it frightened the horses, made travel on the roads unsafe, and was "bad for business generally." As the machine had served its purpose with Curtiss, and had given Hammondsport its little diversion, the famous "wind-wagon" passed into history, and, like so many other of Curtiss' experiments, remains only in the memories of those who were directly interested or those who watched in idle curiosity.

Other airships were built by Baldwin and Curtiss from time to time, and these were used successfully in giving exhibitions throughout the United States. The work of these two pioneers of the air had attracted the attention of the United States Government, in the meantime, and great was the elation at Hammondsport when an order came from the War Department at Washington for a big dirigible balloon for the use of the Signal Corps. Baldwin was commissioned to build the balloon and Curtiss the motor to propel it. This was an important undertaking, and both Baldwin and Curtiss appreciated the fact. It marked the beginning of Governmental and military interest in aeronautics in this country, the possibilities of which were already engaging the attention of the military authorities of Europe. The success of this airship meant much to both men, and Baldwin and Curtiss worked all through the winter of 1904-05 to make it so, Baldwin, meanwhile, having moved to Hammondsport in order to be in touch with the Curtiss factory, where all the mechanical parts of his airships were being made.

In order to meet the specifications drawn up by the War Department, the big airship was required to make a continuous flight of two hours under the power of the motor, and be capable of manœuvring in any direction. Curtiss realised that in order to fill these requirements a new type motor would be needed. He designed and set about building, therefore, a water-cooled motor, something which had not been attempted at the Curtiss factory up to this time, and the success of which marked a long step in advance. Although Baldwin had built thirteen dirigibles, all of which had been equipped with motors built by Curtiss, and all of which had been operated successfully in exhibitions, the Government contract was his most ambitious undertaking. About the balloon itself, there was never any doubt; the thing that clung constantly in the minds of these men who were bending every effort to the conquest of the air, was: "Will the motor do its work in a two-hours' endurance test, and will it furnish the necessary power to drive the big airship at a speed of twenty miles an hour?" The conditions under which the trial was to be made were entirely unique. The motor had to be suspended on a light but substantial framework beneath the great gas-bag, and from this framework the pilot and the engineer had to do their work.

The Army dirigible was completed on time and its test took place at Washington in the summer of 1905. Captain Baldwin acted as pilot and Curtiss as engineer. The airship met every specification and was accepted by the Government. A flight of two hours' duration was made over the wooded hills of Virginia, and this stands to-day as the longest continuous flight ever made by a dirigible airship in this country.

PART II

MY FIRST FLIGHTS

BY
GLENN H. CURTISS

CHAPTER I

BEGINNING TO FLY

IN 1905, while in New York City, I first met Dr. Alexander Graham Bell, the inventor of the telephone. Dr. Bell had learned of our light-weight motors, used with success on the Baldwin dirigibles, and wanted to secure one for use in his experiments with kites. We had a very interesting talk on these experiments, and he asked me to visit him at Bienn Bhreagh, his summer home near Baddeck, Nova Scotia. Dr. Bell had developed some wonderfully light and strong tetrahedral kites which possessed great inherent stability, and he wanted a motor to install in one of them for purposes of experimentation. This kite was a very large one. The Doctor called it an "aerodrome." The surfaces not being planes, it could not properly be described as an aeroplane. He believed that the time would come when the framework of the aeroplane would have to be so large in proportion to its surface that it would be too heavy to fly. Consequently, he evolved the tetrahedral or cellular form of structure, which would allow of the size being increased indefinitely, while the weight would be increased only in the same ratio.

Dr. Bell had invited two young Canadian engineers, F. W. Baldwin and J. A. D. McCurdy, to assist him, and they were at Baddeck when I first visited there in the summer of 1907. Lieutenant Thomas Selfridge, of the United States Army, was also there. Naturally, there was a wide discussion on the subject of aeronautics, and so numerous were the suggestions made and so many theories advanced, that Mrs. Bell suggested the formation of a scientific organisation, to be known as the "Aerial Experiment Association." This met with a prompt and hearty agreement and the association was created very much in the same manner as Dr. Bell had previously formed the "Volta Association" at Washington for developing the phonograph. Mrs. Bell, who was most enthusiastic and helpful, generously offered to furnish the necessary funds for experimental work, and the object of the Association was officially set forth as "to build a practical aeroplane which will carry a man and be driven through the air by its own power."

Dr. Alexander Graham Bell was made chairman; F. W. Baldwin, chief engineer; J. A. D. McCurdy, assistant engineer and treasurer; and Lieut. Thomas Selfridge, secretary; while I was honored with the title of Director of Experiments and Chief Executive Officer. Both Baldwin and

STARTING TO FLY
(A) F. W. Baldwin makes first public flight in America. (B) The "June Bug," June, 1908. (C) Baldwin in Aerial Association's Glider

McCurdy were fresh from Toronto University, where they had graduated as mechanical engineers, and Baldwin later earned the distinction of making the first public flight in a motor-driven, heavier-than-air machine. This was accomplished at Hammondsport, N. Y., March 12, 1908, over the ice on Lake Keuka. The machine used was Number One, built by the Aerial Experiment Association, designed by Lieutenant Selfridge, and known as "The Red Wing." The experiments carried on at Baddeck during the summer and fall of 1907 covered a wide range. There were trials and tests with Dr. Bell's tetrahedral kites, with motors, and with aerial propellers mounted on boats. Finally, at the suggestion of Lieutenant Selfridge, it was decided to move the scene of further experiments to Hammondsport, N. Y., where my factory is located, and there to build a glider. I had preceded the other members of the Association from Baddeck to Hammondsport in order to prepare for the continuance of our work. A few days after my return I was in my office, talking to Mr. Augustus Post, then the Secretary of the Aero Club of America, when a telegram came from Dr. Bell, saying: "Start building. The boys will be down next week." As no plans had been outlined, and nothing definite settled upon in the way of immediate experiments, I was somewhat undecided as to just what to build. We then discussed the subject of gliders for some time and I finally decided that the thing to do was to build a glider at the factory and to take advantage of the very abrupt and convenient hills at Hammondsport to try it out. We therefore built a double-surface glider of the Chanute type.

As almost every schoolboy knows in this day of advanced information on aviation, a glider is, roughly speaking, an aeroplane without a motor. Usually it has practically the same surfaces as a modern aeroplane, and may be made to support a passenger by launching it from the top of a hill in order to give it sufficient impetus to sustain its own weight and that of a rider. If the hill is steep the glider will descend at a smaller angle than the slope of the hill, and thus glides of a considerable distance may be made with ease and comparative safety.

Our first trials of the glider, which we built on the arrival of the members of the Experiment Association, were made in the dead of winter, when the snow lay deep over the hillsides. This made very hard work for everybody. It was a case of trudging laboriously up the steep hillsides and hauling or carrying the glider to the top by slow stages. It was easy enough going down, but slow work going up; but we continued our trials with varied success until we considered ourselves skilful enough to undertake a motor-driven machine, which we mounted on runners.

CHAPTER II

FIRST FLIGHTS

IT was my desire to build a machine and install a motor at once, and thus take advantage of the opportunity furnished by the thick, smooth ice over Lake Keuka at that season of the year. But Lieutenant Selfridge, who had read a great deal about gliders and who had studied them from every angle, believed we should continue experimenting with the glider. However, we decided to build a machine which we believed would fly, and in due time a motor was installed and it was taken down on Lake Keuka to be tried out. We called it the "Red Wing," and to Lieutenant Selfridge belongs the honour of designing it, though all the members of the Aerial Experiment Association had some hand in its construction. We all had our own ideas about the design of this first machine, but to Lieutenant Selfridge was left the privilege of accepting or rejecting the many suggestions made from time to time, in order that greater progress might be made. A number of our suggestions were accepted, and while the machine as completed cannot properly be described as the result of one man's ideas, the honour of being the final arbiter of all the problems of its design certainly belongs to Lieutenant Selfridge.

Now that the machine was completed and the motor installed, we waited for favourable weather to make the first trial. Winter weather around Lake Keuka is a very uncertain element, and we had a long, tiresome wait until the wintry gales that blew out of the north gave way to an intensely cold spell. Our opportunity came on March 12, 1908. There was scarcely a bit of wind, but it was bitterly cold. Unfortunately, Lieutenant Selfridge was absent, having left Hammondsport on business, and "Casey" Baldwin was selected to make the first trial. We were all on edge with eagerness to see what the machine would do. Some of us were confident, others sceptical.

Baldwin climbed into the seat, took the control in hand, and we cranked the motor. When we released our hold of the machine, it sped over the ice like a scared rabbit for two or three hundred feet, and then, much to our joy, it jumped into the air. This was what we had worked for through many long months, and naturally we watched the brief and uncertain course of Baldwin with a good deal of emotion. Rising to a height of six or eight feet, Baldwin flew the unheard-of distance of three hundred and eighteen feet, eleven inches! Then he came down ingloriously on one wing. As we learned afterward, the frail framework of the tail had bent and the machine had flopped over on its side and dropped on the wing, which gave way and caused the machine to turn completely around.

But it had been a successful flight—and we took no toll of the damage to the machine or the cost. We had succeeded! that was the main thing. We had actually flown the "Red Wing" three hundred and eighteen feet and eleven inches! We knew now we could build a machine that would fly longer and come down at the direction of the operator with safety to both.

It had taken just seven weeks to build the machine and to get it ready for the trial; it had taken just about twenty seconds to smash it.

But a great thing had been accomplished. We had achieved the first public flight of a heavier-than-air machine in America!

As our original plans provided for the building of one machine designed by each member of the Association, with the assistance of all the others, the building of the next one fell to Mr. Baldwin, and it was called the "White Wing." The design of the "Red Wing" was followed in many details, but several things were added which we believed would give increased stability and greater flying power. The construction of the "White Wing" was begun at once, but before we could complete it the ice on the lake had yielded to the spring winds and we were therefore obliged to transfer our future trials to land. This required wheels for starting and alighting in the place of the ice runners used on the "Red Wing." An old half-mile race track a short distance up the valley from the Lake was rented and put in shape for flights. The place was called "Stony Brook Farm," and it was for a long time afterward the scene of our flying exploits at Hammondsport.

It would be tiresome to the reader to be told of all the discouragements we met with; of the disheartening smashes we suffered; how almost every time we managed to get the new machine off the ground for brief but encouraging flights, it would come down so hard that something would give way and we would have to set about the task of building it up again. We soon learned that it was comparatively easy to get the machine up in the air, but it was most difficult to get it back to earth without smashing something. The fact was, we had not learned the art of landing an aeroplane with ease and safety—an absolutely necessary art for every successful aviator to know. It seemed one day that the limit of hard luck had been reached, when, after a brief flight and a somewhat rough landing, the machine folded up and sank down on its side, like a wounded bird, just as we were feeling pretty good over a successful landing without breakage.

Changes in the details of the machine were many

and frequent, and after each change there was a flight or an attempted flight. Sometimes we managed to make quite a flight, and others—and more numerous—merely short "jumps" that would land the machine in a potato patch or a cornfield, where, in the yielding ground, the wheels would crumple up and let the whole thing down. Up to this time we had always used silk to cover the planes, but this proved very expensive and we decided to try a substitute. An entirely new set of planes were made and the new covering put on them. They looked very pretty and white as we took the rebuilt machine out with every expectation that it would fly. Great was our surprise, however, when it refused absolutely to make even an encouraging jump. For a time we were at a loss to understand it. Then the reason became as plain as day; we had used cotton to cover the planes, and, being porous, it would not furnish the sustaining power in flight. This was quickly remedied by coating the cotton covering with varnish, rendering it impervious to the air. After that it flew all right. I believe this was the first instance of the use of a liquid filler to coat the surface cloth. It is now used widely, both in this country and in Europe.

We had a great many minor misfortunes with the "White Wing," but each one taught us a lesson. We gradually learned where the stresses and strains lay, and overcame them. Thus, little by little, the machine was reduced in weight, simplified in detail, and finally took on some semblance to the standard Curtiss aeroplane of to-day.

All the members of the Aerial Experiment Association were in Hammondsport at this time, including Dr. Alexander Graham Bell. We had established an office in the annex which had been built on the Curtiss homestead, and here took place nightly discussions on the work of the day past and the plans for the day to follow. Some of the boys named the office the "thinkorium." Every night the minutes of the previous meeting would be read and discussed. These minutes, by the way, were religiously kept by Lieutenant Selfridge and later published in the form of a bulletin and sent to each member. Marvellous in range were the subjects brought up and talked over at these meetings! Dr. Bell was the source of the most unusual suggestions for discussion. Usually these were things he had given a great deal of thought and time to, and, therefore, his opinions on any of his hobbies were most interesting. For instance, he had collected a great deal of information on the genealogy of the Hyde family, comprising some seven thousand individuals. These he had arranged in his card index system, in order to determine the proportion of male and female individuals, their relative length of life, and other characteristics. Or, perhaps, the Doctor would talk about his scheme to influence the sex of sheep by a certain method of feeding; his early experiences with the telephone, the phonograph, the harmonic telegraph, and multiple telegraphy. At other times we would do a jig-saw puzzle with pictures of aeroplanes, or listen to lectures on physical culture by Dr. Alden, of the village. Then, for a change, we would discuss, with great interest and sincerity, the various methods of making sounds to accompany the action of a picture, behind the curtain of the moving-picture show, which we all had attended. Motorcycle construction and operation were studied at the factory and on the roads around Hammondsport. McCurdy used to give us daily demonstrations of how to fall off a motorcycle scientifically. He fell off so often, in fact, that we feared he would never make an aviator. In this opinion, of course, we were very much in error, as he became one of the first, and also one of the best aviators in the country. Atmospheric pressure, the vacuum motor, Dr. Bell's tetrahedral construction, and even astronomical subjects—all found a place in the nightly discussions at the "thinkorium."

Of course there were many important things that took up our attention, but we could not always be grave and dignified. I recall one evening somebody started a discussion on the idea of elevating Trinity Church, in New York City, on the top of a skyscraper, and using the revenue from the ground rental to convert the heathen. This gave a decided shock to a ministerial visitor who happened to be present.

When summer came on there were frequent motorcycle trips when the weather did not permit of flying, or when the shop was at work repairing one of our frequent smashes. "Casey" Baldwin and McCurdy furnished a surprise one day by a rather unusual long-distance trip on motorcycles. "Let's go up to Hamilton, Ontario," said Baldwin, probably choosing Hamilton as the destination because he was charged with having a sweetheart there.

"All right," answered McCurdy.

Without a moment's hesitation the two mounted their wheels, not even stopping to get their caps, and rode through to Hamilton, a hundred and fifty miles distant, buying everything they required along the way. They were gone a week and came back by the same route.

A favourite subject of talk at the "thinkorium," at least between McCurdy and Selfridge, was on some of the effects of the "torque" of a propeller and whenever this arose we would expect the argument to keep up until one or the other would fall asleep.

After the nightly formal sessions of the members of the Association the courtesy of the floor was extended to any one who might be present for the discussion of anything he might see fit to bring up. Later we would adjourn to Dr. Bell's room, where he would put himself into a comfortable position, light his inevitable pipe, and produce his note books. In these note books Dr. Bell would write down everything—his thoughts on every subject imaginable, his ideas about many things, sketches, computations. All these he would sign, date, and have witnessed. It was Dr. Bell's custom to work at night when there were no distracting noises, though there were few of these at Hammondsport even during the daylight hours; at night it is quiet enough for the most exacting victim of insomnia. Dr. Bell often sat up until long after midnight, but he made up for the lost time by sleeping until noon. No one was allowed to wake him for any reason. The rest of us were up early in order to take advantage of the favourable flying conditions during the early morning hours. Dr. Bell had a strong aversion to the ringing of the telephone bell—the great invention for which he is responsible. I occasionally went into his room and found the bell stuffed with paper, or wound around with towels.

"Little did I think when I invented this thing," said Dr. Bell, one day when he had been awakened by the jingling of the bell, "that it would rise up to mock and annoy me."

While the Doctor enjoyed his morning sleep we

were out on "Stony Brook Farm" trying to fly. We had put up a tent against the side of an old sheep barn, and out of this we would haul the machine while the grass was still wet with dew. One never knew what to expect of it. Sometimes a short flight would be made; at others, something would break. Or, maybe, the wind would come up and this would force us to abandon all further trials for the day. Then it was back to the shop to work on some new device, or to repair damages until the wind died out with the setting of the sun. Early in the morning and late in the evening were the best periods of the day for our experimental work because of the absence of wind.

On May 22, 1908, our second machine, the "White Wing," was brought to such a state of perfection that I flew it a distance of one thousand and seventeen feet in nineteen seconds, and landed without damage in a ploughed field outside the old race track. It was regarded as a remarkable flight at that time, and naturally, I felt very much elated.

CHAPTER III

THE "JUNE BUG"—FIRST FLIGHTS FOR THE SCIENTIFIC AMERICAN TROPHY AND FIRST EXPERIMENTS WITH THE HYDROAEROPLANE

FOLLOWING the success of the "White Wing," we started in to build another machine, embodying all that we had learned from our experience with the two previous ones. Following our custom of giving each machine a name to distinguish it from the preceding one, we called this third aeroplane the "June Bug." The name was aptly chosen, for it was a success from the very beginning. Indeed, it flew so well that we soon decided it was good enough to win the trophy which had been offered by *The Scientific American* for the first public flight of one kilometer, or five-eights of a mile, straightaway. This trophy, by the way, was the first to be offered in this country for an aeroplane flight, and the conditions specified that it should become the property of the person winning it three years in succession. The "June Bug" was given a thorough try-out before we made arrangements to fly for the trophy, and we were confident it would fulfill the requirements.

The Fourth of July, 1908, was the day set for the trial. A large delegation of aero-club members came on from New York and Washington, among whom were Stanley Y. Beach, Allan R. Hawley, Augustus Post, David Fairchild, Chas. M. Manley, Christopher J. Lake, A. M. Herring, George H. Guy, E. L. Jones, Wilbur R. Kimball, Captain Thomas S. Baldwin and many other personal friends. The excitement among the citizens of Hammondsport in general was little less than that existing among the members of the Aerial Experiment Association, and seldom had the Fourth of July been awaited with greater impatience.

When Independence Day finally dawned it did not look auspicious for the first official aeroplane flight for a trophy. Clouds boded rain and there was some wind. This did not deter the entire population of Hammondsport from gathering on the heights around the flying field, under the trees in the valley and, in fact, at every point of vantage. Some were on the scene as early as five o'clock in the morning, and many brought along baskets of food and made a picnic of it. The rain came along toward noon, but the crowd hoisted its umbrellas or sought shelter under the trees and stayed on. Late in the afternoon the sky cleared and it began to look as if we were to have the chance to fly after all. The "June Bug" was brought out of its tent and the motor given a try-

THE FIRST MACHINES
(A) "The White Wing," Baldwin driving. 1908. (B) Selfridge's "Red Wing" on the ice, Lake Keuka

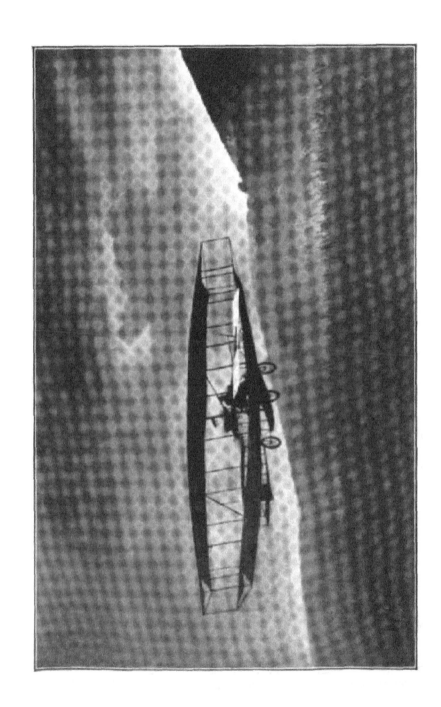

out. It worked all right. The course was measured and a flag put up to mark the end. Everything was ready and about seven o'clock in the evening the motor was started and I climbed into the seat. When I gave the word to "let go" the "June Bug" skimmed along over the old race track for perhaps two hundred feet and then rose gracefully into the air. The crowd set up a hearty cheer, as I was told later—for I could hear nothing but the roar of the motor and I saw nothing except the course and the flag marking a distance of one kilometer. The flag was quickly reached and passed and still I kept the aeroplane up, flying as far as the open fields would permit, and finally coming down safely in a meadow, fully a mile from the starting place. I had thus exceeded the requirements and had won the Scientific American Trophy for the first time. I might have gone a great deal farther, as the motor was working beautifully and I had the machine under perfect control, but to have prolonged the flight would have meant a turn in the air or passing over a number of large trees. The speed of this first official flight was closely computed at thirty-nine miles an hour.

Dr. Bell had gone to Nova Scotia, unfortunately, and, therefore, did not witness the Fourth of July flight of the "June Bug." The other members, however, were all present. It was a great day for all of us and we were more confident than ever that we had evolved, out of our long and costly experiments, a machine that would fly successfully and with safety to the operator. Lieutenant Selfridge was particularly enthusiastic, and I recall when Mr. Holcomb, special agent for a life insurance company, visited the field one day and heard Selfridge talk about flying.

"You must be careful, Selfridge," said Mr. Holcomb, "or we will need a bed for you in the hospital of which I am a trustee."

"Oh, I am careful, all right," replied Selfridge, but it was only a few days later when he left Hammondsport for Washington, and was killed while flying as a passenger with Orville Wright at Fort Meyer.

In Selfridge we lost not only one of the best-posted men in the field of aeronautics, a student and a man of practical ideas, but one of our best-loved companions and co-workers, as well.

Three machines had thus far been built and flown, first the "Red Wing," designed by Lieutenant Selfridge; next the "White Wing," by Baldwin, and last the "June Bug," by me. It was now McCurdy's turn and he designed a machine which he named the "Silver Dart." While this was building we decided to take the "June Bug" down to the lake, equip it with a set of pontoons, or a boat, and attempt to fly from the water. It was my idea that if we could design a float that would sustain the aeroplane on an even keel and at the

THE "JUNE BUG" 55

same time furnish a minimum of resistance, we would be able to get up enough speed to rise from the water. Besides, the lake would afford an ideal flying place, and, what was more important still, a fall or a bad landing would not be nearly so likely to result in injury to the aviator.

Accordingly, we mounted the "June Bug" on two floats, built something like a catamaran, and re-named it the "Loon." It required some time to construct light and strong floats and it was not until the beginning of November, 1908, that we were ready for the first attempt to fly from the water ever made in this or any other country. The "Loon" was hauled down to the lake from the aerodrome on a two-wheeled cart, there being no wheels for rolling it over the ground. I remember we had to build a platform on the cart and to strengthen the wheels to carry the weight of nearly one thousand pounds which the added equipment had brought the total weight up to.

This first experimental hydroaeroplane was a crude affair as compared with the machine in which I made the first successful flight from and landing upon the water, more than three years later at San Diego, Cal. The cleaner lines, the neat, light-weight boat and the other details of the Curtiss hydroaeroplane offer as striking a contrast to the "Loon" as the modern locomotive offers to the crude, clumsy affairs that now exist only in the museums. So great is the difference that one is inclined to marvel that we had any success whatever with the first design.

We made many attempts to rise from the water in the "Loon," but owing to the great weight were unable to make any real flights, although the observers on shore were sure that the pontoons were sometimes clear of the water. By the end of November our experiments had convinced every one of us that we needed more power—and more time than we had at our disposal just then. The best motor we had at our command was able to deliver only enough power to drive the "Loon" at twenty-five miles an hour on the water. This was not enough to get the machine into the air, unless assisted by a strong head wind, and we were not anxious to try flying in a strong wind.

In the meantime McCurdy's machine, the "Silver Dart," had been completed and mounted on wheels. The first flight was made by McCurdy on December 12, 1908, over the "Stony Brook" flying field. The "Silver Dart" was practically the same as the "June Bug." Shortly after this it was shipped to Dr. Bell's place at Baddeck, Nova Scotia, where McCurdy and "Casey" Baldwin used it all through the winter in practice, making flights from the ice and covering all the country thereabouts. McCurdy estimates that in his some two hundred flights in the "Silver Dart," he covered more than a thousand miles.

CHAPTER IV

FIRST FLIGHTS IN NEW YORK CITY

AS a result of the winning of the Scientific American Trophy, the Aeronautical Society of New York City placed an order in the winter of 1908-09 for an aeroplane to be demonstrated at Morris Park Track, New York City, in the spring.

Plans were outlined for enlarging the Hammondsport factory and work commenced on the machine ordered by the Aeronautical Society. It was the plan of this Society to purchase the aeroplane and have one or more of its members taught to fly it. The machine was finished in due time, thoroughly tried out at Hammondsport before it was shipped to New York, and finally sent to the old Morris Park Race Track, where the Aeronautical Society had arranged for the first public exhibition ever held in the history of aviation. There, on June 26, 1909, I had the honour of making the first aeroplane flights in New York City, in the machine bought by the Aeronautical Society.

The Society intended to make Morris Park the scene of aviation meets and of experiments with gliders, but the grounds proved too small and I recommended a change to some other place in the vicinity of New York City, where there was plenty of open country and where the danger from unexpected landings would be minimized. I looked over all the suitable places around New York City and finally decided upon Mineola, on Long Island. The Hempstead Plains, a large, level tract lying just outside Mineola, offered an ideal place for flying and the Aeronautical Society machine was brought down there from Morris Park.

There was such a fine field for flying at Mineola that I decided to make another try for the Scientific American trophy, which I had won on the previous Fourth of July at Hammondsport with the "June Bug." I wanted that trophy very much, but in order to become possessed of it I had to win it three years in succession, the conditions being changed from year to year to keep pace with the progress and development of aviation. The second year's conditions required a continuuous flight of more than twenty-five kilometers (about sixteen miles) in order to have the flight taken into account in awarding the prize, which was to go to the person making the longest official flight during the year.

I believed I could make a fine showing at Hempstead Plains and preparations were made for the attempt. The aeroplane was put together near Peter McLaughlin's hotel and a triangular course

FLIGHTS IN NEW YORK 59

of one and a third miles was measured off. After I had made a number of trial flights over the course I sent formal notice to the Aero Club of America that all was ready for the official flight, and the Club sent Mr. Charles M. Manley down as official representative to observe the trial for the Scientific American trophy.

On July 17th, 1909, a little more than a year from the first official flight of the "June Bug" at Hammondsport, we got out on the field at Mineola at sunrise, before the heavy dew was off the grass, and made ready. It was a memorable day for the residents of that particular section of Long Island, who had never seen a flying machine prior to my brief trial flights there a few days before. They turned out in large numbers, even at that early hour, and there was a big delegation of newspapermen from the New York dailies on hand. Flying was such a novelty at that time that nine-tenths of the people who came to watch the preparations were sceptical while others declared that "that thing won't fly, so what's the use of waiting 'round." There was much excitement, therefore, when, at a quarter after five o'clock, on the morning of July 17, I made my first flight. This was for the Cortlandt Field Bishop prize of two hundred and fifty dollars, offered by the Aero Club of America to the first four persons who should fly one kilometer. It took just two and a half minutes to win this prize and immediately afterward I started for the Scientific American trophy.

The weather was perfect and everything worked smoothly. I made twelve circuits of the course, which completed the twenty-five kilometers, in thirty-two minutes. The motor was working so nicely and the weather man was so favourable, that I decided to keep right on flying, until finally I had circled the course nineteen times and covered a distance of twenty-four and seven-tenths miles before landing. The average speed was probably about thirty-five miles an hour, although no official record of the speed was made.

Great was the enthusiasm of the crowd when the flight ended. I confess that I, too, was enthusiastic over the way the motor had worked and the ease with which the machine could be handled in flight. Best of all, I had the sense of satisfaction that the confidence imposed in me by my friends had been justified.

As the machine built for the Aeronautical Society had thus met every requirement, I agreed to teach two members to fly at Hempstead Plains. Mr. Charles F. Willard and Mr. Williams were the two chosen to take up instruction, and the work began at once. Mr. Willard proved an apt pupil and after a few lessons mastered the machine and flew with confidence and success, circling about the country around Mineola.

FLIGHTS IN NEW YORK

These flights at Mineola gave that place a start as the headquarters for aviators, and it soon became the popular resort for everyone interested in aviation in and near the city of New York.

SCIENTIFIC AMERICAN TROPHY

PART III

MY CHIEF FLIGHTS AND THE WORK OF TO-DAY

BY
GLENN H. CURTISS

CHAPTER I

THE RHEIMS MEET—FIRST INTERNATIONAL AEROPLANE CONTEST

PRIOR to the first flights in New York City I had formulated plans for an improved machine, designed for greater speed and equipped with a more powerful motor. I wanted to take part in the first contest for the Gordon Bennett Aviation cup at Rheims, France, August 22 to 29, 1909. This was the first International Aviation Meet held, and much was expected of the French machines of the monoplane type. Great was my gratification, therefore, when I received word from the Aero Club of America, through Mr. Cortlandt Field Bishop, who was then president, that I had been chosen to represent America at Rheims.[1]

Without allowing my plans to become known to the public I began at once to build an eight-cyl-

---

[1] It is interesting to note that Lieutenant Frank P. Lahm, the sole American entrant for the Gordon Bennett Balloon Cup in 1906; Mr. Edgar Mix, the only representative of America in the balloon contest in 1909, and Mr. Charles Weymann, the only entrant from America in the Gordon Bennett Aviation Cup race of 1911, held in England, all won.

inder, V-shaped, fifty horse-power motor. This was practically double the horse-power I had been using. Work on the motor was pushed day and night at Hammondsport, as I had not an hour to spare. I had kept pretty close watch on everything that had been printed about the preparations of the Frenchmen for the Gordon Bennett race and although it was reported that Blériot, in his own monoplane, and Hubert Latham, in an Antoinette monoplane, had flown as fast as sixty miles an hour, I still felt confident. The speed of aeroplanes is so often exaggerated in press accounts that I did not believe all I read about Blériot's and Latham's trial flights.

The motor was finished, but there was no time to put it in the new machine and try it out before sailing. It was, therefore, given a short run on the block, or testing-frame, hurriedly packed, and the entire equipment rushed to New York barely in time to catch the steamer for France.

The time was so short between the arrival of our steamer and the opening of the meet that in order to get to Rheims in time to qualify, we had to take the aeroplane with us on the train as personal baggage. Thanks to the kindness of the French railway officials, who realised our situation, and evidently had imbibed some of the prevailing aviation enthusiasm, we arrived at Rheims in quick time. In those early days of aviation there was not the keen partisanship for mono-

THE RHEIMS MEET 67

plane or biplane that one finds everywhere to-day; nor was there the strong popular feeling in France in favor of the monoplane that exists to-day. An aeroplane was simply an aeroplane at that time, and interesting as such, but naturally all Frenchmen favored their compatriots who were entered in the race, particularly Blériot, who had just earned world-wide fame by his flight across the English channel. The Frenchmen, as well as Europeans in general, fully expected Blériot to win with his fast monoplane.

My own personal hopes lay in my motor. Judge of my surprise, therefore, upon arriving at Rheims, to learn that Blériot, who had probably heard through newspaper reports that I was bringing over an eight-cylinder motor, had himself installed an eight-cylinder motor of eighty horse-power in one of his light monoplanes. When I learned this, I believed my chances were very slim indeed, if in fact they had not entirely disappeared. The monoplane is generally believed to be faster than the biplane with equal power. I had just one aeroplane and one motor; if I smashed either of these it would be all over with America's chances in the first International Cup Race. I had not the reserve equipment to bring out a new machine as fast as one was smashed, as Blériot and other Frenchmen had. Incidentally, there were many of them smashed during the big meet on the Plain of Bethany. At one time, while flying, I saw as many as twelve machines strewn about the field, some wrecked and some disabled and being hauled slowly back to the hangars, by hand or by horses. For obvious reasons, therefore, I kept out of the duration contests and other events, flying only in such events as were for speed, and of a distance not to exceed twenty kilometers, which was the course for the Gordon Bennett contest in 1909.

It is hard enough for any one to map out a course of action and stick to it, particularly in the face of the desires of one's friends; but it is doubly hard for an aviator to stay on the ground waiting for just the right time to get into the air. It was particularly hard for me to keep out of many events at Rheims held from day to day, especially as there were many patriotic Americans there who would have liked to see America's only representative take part in everything on the programme. I was urged by many of these to go out and contest the Frenchmen for the rich prizes offered and it was hard to refuse to do this. These good friends did not realise the situation. America's chances could not be imperilled for the sake of gratifying one's curiosity, or national pride. On top of the urgings of my American friends to go out and fly and take chances of having a whole machine when the day for the Gordon Bennett should arrive, I was penalised for not starting in the speed race, the Prix de la Vitesse, the penalty being one-twentieth of the time made when I should start in this event. However, I made a number of trial flights and ten official ones, during the meet, without mishap, except a sprained ankle. This was the result of running through growing grain at the time of landing and being thrown out of the machine. I was also fortunate in being the only aviator who took part in this first big meet to land at the hangar after each flight.

During this period of waiting, and making explanations to enthusiastic Americans who could not understand why I did not fly all the time, my mechanician, "Tod" Shriver,[1] attracted a tremendous amount of attention from the throngs that visited the hangars because he worked in his shirt sleeves. They thought "Tod" picturesque because he did not wear the French workman's

[1] Tod Shriver, or "Slim" as he was known to all American aviators because he was very tall and slender, went to Rheims as a mechanic before taking up flying himself. He was successful as an aviator and accompanied Captain Thomas Baldwin to the Orient in the spring and summer of 1911. This trip created great excitement among the Chinese, who had never seen the "foreign devils" fly before. Captain Baldwin tells a story of the crowd that witnessed the flights in Tokyo, Japan, which he describes as numbering seven hundred thousand persons! In proof of this he states that advices received from Japan in the spring of 1912 report that the crowd had not entirely dispersed even at that time! "Tod" Shriver flew in many places in the United States and in the winter of 1911 met his death in Porto Rico. He fell while flying at Ponce. His death was a shock to his many friends.
—[Note by Augustus Post.]

blouse. Shriver used to say that if he were picturesque in shirt sleeves there were about fifty million perfectly good Americans across the Atlantic who formed probably the most picturesque crowd on earth.

In the try-outs it became evident to the Frenchmen that my aeroplane was very fast and it was conceded that the race for the Gordon Bennett Cup would lie between Blériot and myself, barring accidents. After a carefully timed trial circuit of the course, which, much to my surprise, I made in a few seconds less than M. Blériot's time, and that, too, with my motor throttled down slightly, I gained more confidence. I removed the large gasoline tank from my machine and put on a smaller one in order to lessen the weight and the head-resistance. I then selected the best of my three propellers, which, by the way, were objects of curiosity to the French aviators, who were familiar only with the metal blades used on the Antoinette machine, and the Chauvière, which was being used by M. Blériot. M. Chauvière was kind enough to make a propeller especially fitted to my aeroplane, notwithstanding the fact that a better propeller on my machine would lessen the chances of the French flyers for the cup. However, I decided later to use my own propeller, and did use it—and won.

August 29 dawned clear and hot. It was agreed

THE RHEIMS MEET 71

at a meeting of the Committee, at which all the contestants were present, that each contestant should be allowed to make one trial flight over the course and that he might choose his own time for making it, between the hours of ten o'clock in the morning and six o'clock in the evening. The other starters were Blériot, Lefebre, and Latham for France, and Cockburn for England. As I have already stated, Blériot was the favourite because of his trip across the English channel and because of his records made in flights at various places prior to the Rheims meet.

As conditions were apparently good, I decided to make my trial flight shortly after ten o'clock. The machine was brought out, the engine given a preliminary run, and at half past ten I was in the air. Everything had looked good from the ground, but after the first turn of the course I began to pitch violently. This was caused by the heat waves rising and falling as the cooler air rushed in. The up and down motion was not at all pleasant and I confess that I eased off on the throttle several times on the first circuit. I had not then become accustomed to the feeling an aviator gets when the machine takes a sudden drop. On the second round I got my nerve back and pulled the throttle wide open and kept it open. This accounts for the fact that the second lap was made in faster time than the first. The two circuits were made safely and I crossed the finish line in seven minutes, fifty-five seconds, a new record for the course.

Now was my chance! I felt that the time to make the start for the Cup was then, in spite of the boiling air conditions, which I had found existed all over the course and made flying difficult if not actually dangerous. We hurriedly refilled the gasoline tank, sent official notice to the judges, carefully tested the wiring of the machine by lifting it at the corners, spun the propeller, and the official trial was on. I climbed as high as I thought I might without protest, before crossing the starting line—probably five hundred feet—so that I might take advantage of a gradual descent throughout the race, and thus gain additional speed. The sun was hot and the air rough, but I had resolved to keep the throttle wide open. I cut the corner as close as I dared and banked the machine high on the turns. I remember I caused great commotion among a big flock of birds which did not seem to be able to get out of the wash of my propeller. In front of the tribunes the machine flew steadily, but when I got around on the back stretch, as we would call it, I found remarkable air conditions. There was no wind, but the air seemed fairly to boil. The machine pitched considerably, and when I passed above the "graveyard," where so many machines had gone down and were smashed during the previous days of the meet, the air seemed literally to drop from under me. It was so bad at one spot that I made up my mind that if I got over it safely I would avoid that particular spot thereafter.

Finally, however, I finished the twenty kilometers in safety and crossed the line in fifteen minutes, fifty seconds, having averaged forty-six and one-half miles an hour. When the time was announced there was great enthusiasm among the Americans present, and every one rushed over to offer congratulations. Some of them thought that I would surely be the winner, but of this I was by no means certain. I had great respect for Blériot's ability, and besides, Latham and his Antoinette might be able to make better speed than they had thus far shown. In a contest of this sort it is never safe to cheer until all the returns are in. I confess that I felt a good deal like a prisoner awaiting the decision of a jury. I had done my best, and had got the limit of speed out of the machine; still I felt that if I could do it all over again I would be able to improve on the time. Meantime Cockburn, for England, had made a start but had come down and run into a haystack. He was only able to finish the course in twenty minutes, forty-seven and three-fifth seconds. This put him out of the contest.

Latham made his trial during the afternoon

but his speed was five or six miles an hour slower than my record. The other contestants were flying about thirty-five miles an hour, and were, therefore, not really serious factors in the race.

It was all up to M. Blériot. All day long he tinkered and tested, first with one machine and then another; trying different propellers and making changes here and there. It was not until late in the afternoon that he brought out his big machine, Number 22, equipped with an eight-cylinder water-cooled motor, mounted beneath the planes, and driving by chain a four-bladed propeller, geared to run at a speed somewhat less than that of the engine. He started off at what seemed to be a terrific burst of speed. It looked to me just then as if he must be going twice as fast as my machine had flown; but it must be remembered that I was very anxious to have him go slow. The fear that he was beating me was father to the belief.

As soon as Blériot was off Mr. Cortlandt Field Bishop and Mr. David Wolfe Bishop, his brother, took me in their automobile over to the judges' stand. Blériot made the first lap in faster time than I had made it, and our hearts sank. Then and there I resolved that if we lost the cup I would build a faster aeroplane and come back next year to win it.

Again Blériot dashed past the stand and it seemed to me that he was going even faster than

WINNING THE GORDON BENNET CONTEST IN FRANCE
(A) Curtiss flying at Rheims. (B) The welcome home to Hammondsport.

"A POSITION HIGHER THAN THE PRESIDENT'S"
President Taft watching Curtiss fly. Harvard Meet. 1910

the first time. Great was my surprise, therefore, when, as he landed, there was no outburst of cheers from the great crowd. I had expected a scene of wild enthusiasm, but there was nothing of the sort. I sat in Mr. Bishop's automobile a short distance from the judges' stand, wondering why there was no shouting, when I was startled by a shout of joy from my friend, Mr. Bishop, who had gone over to the judges' stand.

"You win! You win!" he cried, all excitement as he ran toward the automobile. "Blériot is beaten by six seconds!"

A few moments later, just at half past five o'clock, the Stars and Stripes were slowly hoisted to the top of the flagpole and we stood uncovered while the flag went up. There was scarcely a response from the crowded grand stands; no true Frenchman had the heart to cheer. A good, hearty cheer requires more than mere politeness. But every American there made enough noise for ten ordinary people, so that numbers really counted for very little in the deep feeling of satisfaction at the result of the first great contest in the history of aviation. Mr. Andrew D. White, accompanied by Mrs. Roosevelt and Miss Ethel Roosevelt, came over to our car and congratulated me. Quentin Roosevelt, who had been in a state of excitement throughout the day, declared it "bully," while his brother Archie wanted to be shown all about the working of the machine. M.

Blériot himself, good sportsman that he is, was among the first to extend congratulations to America and to me personally.

There was a reason beyond the mere patriotism why the Americans felt so happy over the result; it meant that the next international race would be held in the United States, and that the best foreign machines would have to come across the ocean to make a try for the cup the following year.

In commenting upon the result the Paris Edition of the New York *Herald* said that the race had rehabilitated the biplane; that while the lightness and bird-like lines of the monoplane had appealed to the crowd as the ideal representation of artificial flight, "the American aviator proved that the biplane not only possessed qualities of carrying weight and undoubtedly of superior stability, but that, if need be, it can develop speed equal to, if not superior to, its smaller rival."

Offers of engagements to fly in Germany and Italy came pouring in. To accept these meant a good deal of money in prizes, for it had been proven that I had the fastest aeroplane in the world. I accepted some of them, as I had learned that the conditions for flying at the big meets in Europe were almost ideal and that there was a tremendous amount of interest everywhere, among all classes. A big meet was organized at Brescia, Italy, and I went there from Rheims.

THE RHEIMS MEET

Here I carried my first passenger, the celebrated Italian poet and author, Gabriele D'Annunzio. He was wildly enthusiastic over his experience, and upon being brought back to earth said with all the emotion of his people: "Until now I have never really lived! Life on earth is a creeping, crawling business. It is in the air that one feels the glory of being a man and of conquering the elements. There is the exquisite smoothness of motion and the joy of gliding through space— It is wonderful! Can I not express it in poetry? I might try."

And he did express it in poetry, a beautiful work published sometime later.

After winning the Grand Prize at Brescia and taking a wonderful motor trip over the Alps with Mr. Bishop, I hurried home to America to look after my business affairs, about which I had not had time even to think during the Rheims and Brescia meets.

NOTE BY AUGUSTUS POST

Delegations of enthusiastic friends met Mr. Curtiss in New York, among them members of the Aero Club of America and other representative organisations. There followed a series of luncheons and dinners which seemed without end. Among all these the luncheon given by the Aero Club of America at the Lawyers' Club was notable because every one present showed such a warm interest in the success of American aeronautics, and such a firm determination not only to keep the trophy in this country, but to defend it the next year in an aviation meet that should be even greater than that with which Rheims had led the way.

But the real celebration took place in the little village of Hammondsport, the place where Mr. Curtiss was born and reared, and where he knew every man, woman, and child. The men in the factory and all his other warm friends got together and decided that there must be something out of the ordinary when he got back to town. They planned a procession all the way from Bath to Hammondsport, a distance of ten miles, with fireworks along the route. But a heavy rain came on just in time to spoil the fireworks plan, so they engaged a special train and this passed through a glow of red fire all the way home from Bath. At the Hammondsport station there was a carriage to draw him up the hill to his home, and fifty men furnished the motive power. There were arches with "Welcome" in electric lights, banners, fireworks, and speeches. Through the pouring rain there was a continuous procession of his friends and acquaintances—townspeople who had always given him their loyal support and the men from the shop who had made his success possible.

It was after eleven o'clock when the crowd dispersed—an almost unholy hour for Hammondsport.—AUGUSTUS POST.

CHAPTER II

HUDSON-FULTON CELEBRATION—FIRST AMERICAN INTERNATIONAL MEET, AT LOS ANGELES

I WAS not permitted to remain long in Hammondsport, although there was much work for me to do there in the way of planning improvements in the factory, as well as on my aeroplane, which had now come to be known throughout the world by reason of winning the Gordon Bennett Cup. There were tempting offers from all quarters to give exhibitions with the flying machine, which up to that time had been seen in but few places in this country. Some of these offers were accepted because I could not afford to reject them. Moreover, it required a great deal of money to run the shop, and there was no commercial demand for aeroplanes. They were, as yet, valuable only as "show machines," to see which the public was willing to pay goodly sums.

For a long time preparations had been going on at New York City to celebrate the tri-centenary of the discovery of the Hudson river, and the centenary of the first steamboat trip on that stream by Fulton in the *Clermont*. It had been the idea of the originators of the Hudson-Fulton celebration—an idea that was expressed in the tentative plans published long before the celebration itself—that the new conquest of the air should be recognised, in some way, at the same time. At first it was intended that some sort of airship should accompany the naval parade the entire length of the Hudson, with a replica of Hendrik Hudson's *Half Moon* leading the way, Robert Fulton's old steamboat *Clermont* following, and the airship hovering above them—thus furnishing a striking illustration of the wonderful advancement in the means of locomotion in a hundred years, and signalising the new science of air navigation. With this end in view the Celebration Committee engaged the Wright Brothers and myself to bring aeroplanes to New York, furnishing us with every facility on Governor's Island, in the Lower Bay, from which point all flights were to be made.

But aerial navigation in the fall of 1909 was not such a sure and certain thing as all that. Much depended upon the wind and weather, and it was soon demonstrated that the best that could be hoped for at the time of the celebration would be flights made at such times as the wind would permit. Day after day the public waited anxiously for flights to be made up the Hudson from Governor's Island, but day after day the wind blew up or down the Hudson in such blasts that it was not deemed safe to attempt a trip. For it must be remembered that there is scarcely a more difficult course anywhere in the country than over the Hudson river in the vicinity of New York. On both sides of the river, which is a swift-running stream, rise lofty hills, and at some places precipitous cliffs called the Palisades. On the New York side are miles upon miles of lofty apartment houses along Riverside Drive. If the wind blows across the river, either from the east or west, dangerous currents and eddies suck down through the cañon-like streets, or over the steep Palisades, making flying extremely hazardous. For this reason there has never, even up to this time (August, 1912), been any flying to speak of over the Hudson, and for these reasons, the great river will not become a popular flying course for aeroplanes until they are so constructed as to be able to defy the treacherous, puffy wind currents. The hydroaeroplane, however, may navigate the course with safety, as it is perfectly safe in one of these machines to fly within a few feet of the water where there is the least danger from contrary air currents.

So much was printed in the New York newspapers while we were waiting for propitious weather that the public was keyed up to expect great things from the aeroplanes—far greater than the aeroplane could accomplish. Bulletins were posted by the newspapers from day to day, informing the public that flights would surely be made "to-day"—provided the wind abated. In the meantime interest was doubly stimulated by the announcement of a ten-thousand-dollar prize for the first air-flight over Fulton's course, from New York to Albany, or from Albany to New York. One of the paintings made at the time as an "advance notice," I remember, showed so many aerial dreadnaughts in the sky, passing down the river by the Palisades at the same time, that one was forced to wonder how all of them were going to find room to navigate. However, the atmosphere had cleared long before the actual flight was made down the Hudson, the following summer.

In spite of the disappointment felt by the public at not seeing a fleet of aeroplanes sporting over the Hudson daily during the Hudson-Fulton celebration, there were many other things to divert the attention of New York's five millions and some few hundred thousands of visitors from this and other countries. The week of pomp and pageantry culminated in the most wonderful marine and land parades ever staged in this country, and seldom, if ever, excelled in the Old World. The marine parade extended all the way up to Albany, and at every stopping place there was a repetition, on a smaller scale, of the scenes of enthusiasm and general holiday spirit that had prevailed in the Metropolis. New York City was decorated as no one had ever seen it decorated before, and the great fleet of over a hundred warships that swung at anchor in the Hudson were visited by thousands by day and were outlined in myriads of electric lights at night, disguising their ominous guns in soft shadow and giving them a peaceful and almost fairy-like appearance.

Then there were the dirigible balloons to command the attention of the crowds that thronged Riverside Drive waiting for the aeroplanes. They, too, were after the rich prize offered by the *New York World*. They furnished the only real contest during the Hudson-Fulton celebration. There were two of them, one entered by the intrepid Captain Thomas Baldwin, and the other by a Mr. Tomlinson. These were housed in great tents raised within an enclosure at Riverside Drive and One Hundred and Nineteenth street, behind a high fence, on which was painted "Hudson-Fulton Flights." This was the center of interest for great crowds for days during the period of waiting. Captain Baldwin, always popular with the people wherever he goes, was the centre of interest with the crowds that stood around the sheds, watching the mild, blunt noses of the big dirigibles as they bobbed and swayed with the gusts that swept around Grant's Tomb, reminding one of the ceaseless weaving of a restless elephant. But the elements seemed to be as much against the dirigibles as against the aeroplanes. Tomlinson made a start, after a long wait, but came to grief almost at once, while Captain Baldwin fared but little better. His trip extended but a few miles up the river, when he was forced to come down, thus ending the chances of the dirigibles.

The aeroplanes were scarcely more fortunate. October winds around New York are most unruly things, and at that particular period seemed worse than usual. Weather-wise folk learned after awhile to look out at the flags on the high buildings; if they stood out straight from the staff, the people went about their business, knowing there would be no flying that day. But every one kept an ear cocked for the firing of a big cannon on Governor's Island, the signal that a flight was about to be made. Even these were deceiving, for there were so many salutes being fired by the great fleets in the river and bay, that no one could tell when to give heed to gun signals. So the crowds sat along Riverside Drive, or depended upon the unhappy and over-worked policemen for word of the aeroplanes. Some people were disposed to hold the policemen personally responsible for the failure of the airships to fly. "You'd think," said one of the blue-coated guardians on Riverside Drive, "that *I* was keepin' 'em back, the way these people go at me. They blame me and not the wind!"

The wind held out and the week of festivities

ended; still there had been no flying. I could not remain in New York any longer, as I had accepted an engagement some time before to fly at St. Louis. I was obliged therefore, much to my chagrin, and the disappointment of the crowds, to leave the city without making a flight up the river, although I did make a short flight over Governor's Island.

Mr. Wilbur Wright, however, remained in New York, and during the following week made a magnificent flight up the river from Governor's Island to Grant's Tomb and return, a distance of about twenty miles. This gave the larger part of New York's millions their first glimpse of an aeroplane in flight.

At St. Louis we gave a very successful meet. There were flights by Captain Baldwin, Lincoln Beachey, and Roy Knabenshue, in their dirigible balloons, and myself in my aeroplane. The weather conditions were favourable, and St. Louis turned out enthusiastic throngs to witness the exhibitions.

The Pacific Coast, always progressive and quick to seize upon every innovation, no matter where it may be developed, had been clamoring for some time for an aviation meet. The enterprising citizens of Los Angeles got together and put up a large sum of money to bring out from Europe and the eastern part of the United States, a number of representative aviators for an interna-

THE LOS ANGELES MEET 87

tional meet, the first ever given in this country. Louis Paulhan, one of the most celebrated French aviators, was brought over with a biplane and a monoplane, and there were a number of American entries, including Charles F. Willard and myself. Los Angeles furnished the first opportunity for a real contest in this country between the French and American machines, and these contests aroused immense interest throughout the country.

The importance of the Los Angeles meet to the aviation industry in this country was very great. The favourable climatic conditions gave opportunities for every one to fly in all the events, and the wide publicity given to the achievements of Paulhan and others, especially to the new world's altitude record established by the French aviator, stimulated interest throughout the country. There was cross-country flying such as had not been seen in this country, brilliant exhibitions of altitude flying, and speed contests of the hair-raising variety. Sometimes it takes just such a public demonstration as the Los Angeles meet not only to spread the news of the general progress of mechanical flight, but to show the builders of aeroplanes themselves just what their machines are capable of.

It was at the Los Angeles meet, by the way, that Charles F. Willard coined that apt and picturesque phrase which soon was used the world over in describing air conditions. Willard had made a short flight and on coming down declared the air "was as full of holes as a Swiss cheese." This made a great hit with the newspapermen, who featured it, using it day after day in their stories until it went the rounds of the press of the world. There were special articles written on "holes in the air," and interviews of prominent aviators to determine how it feels to fall into "a hole in the air."

The expression was more picturesque than accurate, for it is not necessary to explain, in this advanced stage of aviation, that there are no "holes" in the atmosphere. If there were a hole in the atmosphere, a clap of thunder would result, caused by the rushing in of the surrounding air to fill the vacuum. The only holes in the air are the streaks that follow a rifle bullet or a flash of lightning. The real cause of the conditions described by Willard, and which has since probably been responsible for the death of several well known aviators, is a swift, downward current of air, rushing in to fill a vacuum that follows a rising current from a heated area. The hot air rises and the cool air rushes down to take its place. An aeroplane striking one of these descending currents drops as if the entire atmospheric support had been suddenly removed, and if it be not high enough, may strike the ground with fatal results to the aviator. Every experi-

THE LOS ANGELES MEET 89

enced airman has met these conditions. They are especially noticeable over water, streaks of calm water showing where the up-currents are just starting, and waves or ripples where the down-currents strike the surface.

The representative of the Aero Club of America at the Los Angeles meet was Mr. Cortlandt Field Bishop, of New York, who had been at Rheims the previous summer when I won the Gordon Bennett Cup and who had been of inestimable assistance to me at that time. Mr. Bishop had his oft-expressed wish to fly gratified at Los Angeles. He was taken up by Louis Paulhan several times, and Paulhan also took Mrs. Bishop for her first aerial ride. Great crowds came out at the Los Angeles meet, and they for the first time in the history of aviation in this country expected the aviator to fly and not to fall. Paulhan did some wonderful cross-country flying, and as a climax to the week of aerial wonders, he established a world's altitude record by ascending 4,165 feet. This was regarded as marvellous at that time. Since then the mark has been successively raised by Brookins, Hoxsey, Le Blanc, Beachey, Garros and others. Legagneux now (September, 1912) holds the record at 18,760 feet.

Interest in aviation was keen following the Los Angeles meet and I decided to try for the *New York World's* ten-thousand-dollar prize, which was still open, for a flight down the Hudson from

Albany to New York City. Notwithstanding all the natural obstacles in the way of the accomplishment of the undertaking, the conditions were so fair as to stops, time-limit, etc., and it was so obviously a prize offered to be won, that I considered it worth a serious effort.

I fully realised that the flight was much greater than anything I had yet attempted, and even more difficult than Blériot's great flight across the English channel from France to England, news of which was still ringing throughout the world, and even greater than the projected flight from London to Manchester, England, and for which a prize of fifty thousand dollars had been offered. Although the course covered about the same distance as the London-Manchester route, there was not the difficulty of landing safely, over the English route. The Hudson flight meant one hundred and fifty-two miles over a broad, swift stream, flowing between high hills or rugged mountains the entire distance and with seldom a place to land; it meant a fight against treacherous and varying wind currents rushing out unawares through clefts in the mountains, and possible motor trouble that would land both machine and aviator in the water with not much chance of escape from drowning, even if uninjured in alighting.

CHAPTER III

FLIGHT DOWN THE HUDSON RIVER FROM ALBANY TO NEW YORK CITY

TO fly from Albany to New York City was quite an undertaking in the summer of 1910. I realised that success would depend upon a dependable motor and a reliable aeroplane. In preparation for the task, therefore, I set the factory at Hammondsport to work to build a new machine. While awaiting the completion of the machine, I took a trip up the Hudson from New York to Albany to look over the course and to select a place about half way between the two cities where a landing for gasoline and oil might be made, should it become necessary.

There are very few places for an aeroplane to land with safety around New York City. The official final landing place, stipulated in the conditions drawn up by the *New York World*, was to be Governor's Island, but I wanted to know of another place on the upper edge of the city where I might come down if it should prove necessary. I looked all over the upper end of Manhattan Island, and at last found a little meadow on a side hill just at the junction of the Hudson and Harlem rivers, at a place called Inwood. It was small and sloping, but had the advantage of being within the limits of New York City. It proved fortunate for me that I had selected this place, for it later served to a mighty good advantage.

There was quite a party of us aboard the Hudson river boat leaving New York City one day in May for the trip to Albany. As an illustration of the scepticism among the steamboat men, I remember that I approached an officer and asked several questions about the weather conditions on the river, and particularly as to the prevailing winds at that period of the year. Incidentally, I remarked that I was contemplating a trip up the river from New York to Albany in an aeroplane and wanted to collect all the reliable data possible on atmospheric conditions. This officer, whom I afterward learned was the first mate, answered all my questions courteously, but it was evident to all of us that he believed I was crazy. He took me to the captain of the big river boat and introduced me, saying: "Captain, this is Mr. Curtiss, the flying machine man; that's all I know," in a tone that clearly indicated that he disclaimed all responsibility as to anything I might do or say.

The captain was very kind and courteous, asking us to remain in the pilot house, where we might get a better view of the country along the way, and displaying the keenest interest in the project. He answered all our questions about

Copyright, 1910, by The Pictorial News Co.
THE ALBANY-NEW YORK HUDSON FLIGHT
(A) Start of the flight at Albany. Mrs. Curtiss and Augustus Post standing

Copyright, 1910, by The Pictorial News Co.
THE HUDSON FLIGHT
Over Storm King

the winds along the Hudson and seemed to enter heartily in the spirit of the thing until we approached the great bridge at Poughkeepsie and I began to deliberate whether it would be better to pass over or beneath it in the aeroplane. Then it seemed really to dawn upon the captain for the first time that I was actually going to fly down the river in an aeroplane. He apparently failed to grasp the situation, and thereafter his answers were vague and given without interest. It was "Oh, yes, I guess so," and similar doubtful expressions, but when we finally left the boat at Albany he very kindly wished me a safe trip and promised to blow the whistle if I should pass his boat.

Albany afforded a better starting place than New York, because there were convenient spots where one might land before getting well under way, should it become necessary. This was not true of the situation at New York City. As to the advantage of prevailing winds, it seemed to be in favour of Albany as the starting place, and I finally decided to have everything sent up to the capital city. On my way up I had stopped at Poughkeepsie, in order to select a landing place, as at least one stop was deemed necessary to take on gasoline and to look over the motor. We visited the State Hospital for the Insane, which stands on the hill just above Poughkeepsie, and which seemed to be a good place to land. Dr.

Taylor, the superintendent, showed us about the grounds, and when told that I intended stopping there on my way down the river in a flying machine, said with much cordiality: "Why, certainly, Mr. Curtiss, come right in here; here's where all the flying machine inventors land."

Notwithstanding the Doctor's cordial invitation to "drop in on him," we went to the other side of Poughkeepsie, and there found a fine open field at a place called Camelot. I looked over the ground carefully, locating the ditches and furrows, and selected the very best place to make a safe landing. Arrangements were made for a supply of gasoline, water, and oil to be brought to the field and held in readiness. It was fortunate that I looked over the Camelot field, for a few days later I landed within a few feet of the place I had selected as the most favoured spot near Poughkeepsie. This is but one thing that illustrates how the whole trip was outlined before the start was made, and how this plan was followed out according to arrangement.

I shall always remember Albany as the starting place of my first long cross-country flight. My machine was brought over from Hammondsport and set up; the Aero Club sent up its official representatives, Mr. Augustus Post and Mr. Jacob L. Ten Eyck, and the newspapers of New York City sent a horde of reporters. A special train was engaged to start from Albany as soon as I

THE HUDSON FLIGHT 95

got under way, carrying the newspapermen and the Aero Club representatives, as well as several invited guests. It was the purpose to have this train keep even with me along the entire trip of one hundred and fifty-two miles, but as it turned out, it had some trouble in living up to the schedule.

The aeroplane, christened the "Hudson Flier," was set up on Rensselaer Island. It was now up to the weather man to furnish conditions I considered suitable. This proved a hard task, and for three days I got up at daybreak, when there is normally the least wind, ready to make an early start. On these days the newspapermen and officials, not to mention crowds of curious spectators, rubbed the sleep out of their eyes before the sun got up and went out to Rensselaer Island. But the wind was there ahead of us and it blew all day long. The weather bureau promised repeatedly, "fair weather, with light winds," but couldn't live up to promises. I put in some of the time in going over every nut, bolt, and turnbuckle on the machine with shellac. Nothing was overlooked; everything was made secure. I had confidence in the machine. I knew I could land on the water if it became necessary, as I had affixed two light pontoons to the lower plane, one on either end, and a hydro-surface under the front wheel of the landing-gear. This would keep me afloat some time should I come down in the river.

We bothered the life out of the weather observer at Albany, but he was always very kind and took pains to get weather reports from every point along the river. But the newspapermen lost faith; they were tired of the delay. I have always observed that newspapermen, who work at a high tension, cannot endure delay when there is a good piece of news in prospect. One of those at Albany during the wait, offered to lay odds with the others that I would not make a start. Others among the journalists believed I was looking for free advertising, and when another of the advertised starters for the *World* prize reached Albany he was greeted with: "Hello, old man, are you up here to get some free advertising, too?" One of the Poughkeepsie papers printed an editorial about this time, in which it said: "Curtiss gives us a pain in the neck. All those who are waiting to see him go down the river are wasting their time." This was a fair sample of the lack of faith in the undertaking.

The machine was the centre of interest at Albany during the wait. It seemed to hold a fascination for the crowds that came over to the island. One young fellow gazed at it so long and so intently that he finally fell over backwards insensible and it was some time before he was restored to consciousness. Then one of the newspapermen dashed a pail of water over him and at once sent his paper a column about it. They had

THE HUDSON FLIGHT

to find something to write about and the countryman, the flying machine, and the fit made a combination good enough for almost any newspaperman to weave an interesting yarn about.

Our period of waiting almost ended on Saturday morning, May 30th. The "Hudson Flier" was brought out of its tent, groomed and fit; the special train provided by the *New York Times* to follow me over the New York Central, stood ready, with steam up and the engineer holding a right-of-way order through to New York. The newspapermen, always on the job, and the guests were watching eagerly for the aeroplane to start and set out on its long and hazardous flight.

Then something happened—the wind came up. At first it did not seem to be more than a breeze, but it grew stronger and reports from down the river told of a strong wind blowing up the river. This would have meant a head gale all the way to New York, should I make a start then. Everything was called off for the day and we all went over and visited the State Capitol. The newspapermen swallowed their disappointment and hoped for better things on the morrow.

Sunday proved to be the day. The delay had got somewhat on my nerves and I had determined to make a start if there was half a chance. The morning was calm and bright—a perfect summer day. News from down the river was all favourable. I determined it was now or never. I sent

Mrs. Curtiss to the special train and informed the *World* representative and the Aero Club officials that I was ready to go. Shortly after eight o'clock the motor was turned over and I was off!

It was plain sailing after I got up and away from Rensselaer Island. The air was calm and I felt an immense sense of relief. The motor sounded like music and the machine handled perfectly. I was soon over the river and when I looked down I could see deep down beneath the surface. This is one of the peculiar things about flying over the water. When high up a person is able to see farther beneath the surface.

I kept a close lookout for the special train, which could not get under way as quickly as I had, and pretty soon I caught sight of it whirling along on the tracks next to the river bank. I veered over toward the train and flew along even with the locomotive for miles. I could see the people with their heads out the windows, some of them waving their hats or hands, while the ladies shook their handkerchiefs or veils frantically. It was no effort at all to keep up with the train, which was making fifty miles an hour. It was like a real race and I enjoyed the contest more than anything else during the flight. At times I would gain as the train swung around a short curve and thus lost ground, while I continued on in an air line.

All along the river, wherever there was a village or town, and even along the roads and in boats on the river, I caught glimpses of crowds or groups of people with their faces turned skyward, their attitudes betokening the amazement which could not be read in their faces at that distance. Boatmen on the river swung their caps in mute greeting, while now and then a river tug with a long line of scows in tow, sent greetings in a blast of white steam, indicating there was the sound of a whistle behind. But I heard nothing but the steady, even roar of the motor in perfect rhythm, and the whirr of the propeller. Not even the noise of the speeding special train only a few hundred feet below reached me, although I could see every turn of the great drive-wheels on the engine.

On we sped, the train and the aeroplane, representing a century of the history of transportation, keeping abreast until Hudson had been past. Here the aeroplane began to gain, and as the train took a wide sweeping curve away from the bank of the river, I increased the lead perceptibly, and soon lost sight of the special.

It seemed but a few minutes until the great bridge spanning the Hudson at Poughkeepsie, came into view. It was a welcome landmark, for I knew that I had covered more than half the journey from Albany to New York, and that I must stop to replenish the gasoline. I might have gone on and taken a chance on having enough fuel, but this was not the time for taking chances. There was too much at stake.

I steered straight for the centre of the Poughkeepsie bridge, and passed a hundred and fifty feet above it. The entire population of Poughkeepsie had turned out, apparently, and resembled swarms of busy ants, running here and there, waving their hats and hands. I kept close watch for the place where I had planned to turn off the river course and make a landing. A small pier jutting out into the river was the mark I had chosen beforehand and it soon came into view. I made a wide circle and turned inland, over a clump of trees, and landed on the spot I had chosen on my way up to Albany. But the gasoline and oil which I had expected to find waiting for me, were not there. I saw no one for a time, but soon a number of men came running across the fields and a number of automobiles turned off the road and raced toward the aeroplane. I asked for some gasoline and an automobile hurried away to bring it.

I could scarcely hear and there was a continual ringing in my ears. This was the effect of the roaring motor, and strange to say, this did not cease until the motor was started again. From that time on there was no disagreeable sensation. The special train reached the Camelot field shortly after I landed and soon the newspapermen, the Aero Club officials, and the guests came

THE HUDSON FLIGHT 101

climbing up the hill from the river, all eager to extend their congratulations. Henry Kleckler, acting as my mechanic, who had come along on the special train, looked over the machine carefully, testing every wire, testing the motor out, and taking every precaution to make the remainder of the journey as successful as the first half. The gasoline having arrived, and the tank being refilled, the special train got under way; once more I rose into the air, and the final lap of the journey was on.

Out over the trees to the river I set my course, and when I was about midstream, turned south. At the start I climbed high above the river, and then dropped down close to the water. I wanted to feel out the air currents, believing that I would be more likely to find steady air conditions near the water. I was mistaken in this, however, and soon got up several hundred feet and maintained about an even altitude of from five hundred to seven hundred feet. Everything went along smoothly until I came within sight of West Point. Here the wind was nasty and shook me up considerably. Gusts shot out from the rifts between the mountains and made extremely rough riding. The worst spot was encountered between Storm King and Dunderberg, where the river is narrow and the mountains rise abruptly from the water's edge to more than a thousand feet on either side. Here I ran into a downward suction that dropped me in what seemed an interminable fall straight down, but which as a matter of fact was not more than a hundred feet or perhaps less. It was one of Willard's famous "holes in the air." The atmosphere seemed to tumble about like water rushing through a narrow gorge. At another point, a little farther along, and after I had dropped down close to the water, one blast tipped a wing dangerously high, and I almost touched the water. I thought for an instant that my trip was about to end, and made a quick mental calculation as to the length of time it would take a boat to reach me after I should drop into the water.

The danger passed as quickly as it had come, however, and the machine righted itself and kept on. Down by the Palisades we soared, rising above the steep cliffs that wall the stream on the west side. Whenever I could give my attention to things other than the machine, I kept watch for the special train. Now and then I caught glimpses of it whirling along the bank of the river, but for the greater part of the way I outdistanced it.

Soon I caught sight of some of the sky-scrapers that make the sky-line of New York City the most wonderful in the world. First I saw the tall frame of the Metropolitan Tower, and then the lofty Singer building. These landmarks looked mighty good to me, for I knew that, given a few more minutes' time, I would finish the flight. Approaching Spuyten Duyvil, just above the Harlem river, I looked at my oil gauge and discovered that the supply was almost exhausted. I dared not risk going on to Governor's Island, some fifteen miles farther, for once past the Harlem river there would be no place to land short of the island. So I took a wide sweep across to the Jersey side of the river, circled around toward the New York side, and put in over the Harlem river, looking for the little meadow at Inwood which I had picked out as a possible landing place some two weeks before.

There I landed on the sloping hillside, and went immediately to a telephone to call up the *New York World*. I told them I had landed within the city limits and was coming down the river to Governor's Island soon.

I got more oil, some one among the crowd, that gathered as if by magic, turned my propeller, and I got away safely on the last leg of the flight. While I had complied with the conditions governing the flight by landing in the city limits, I wanted to go on to Governor's Island and give the people the chance to see the machine in flight.

From the extreme northern limits of New York to Governor's Island, at the southern limits, was the most inspiring part of the trip. News of the approach of the aeroplane had spread throughout the city, and I could see crowds everywhere.

New York can turn out a million people probably quicker than any other place on earth, and it certainly looked as though half of the population was along Riverside Drive or on top of the thousands of apartment houses that stretch for miles along the river. Every craft on the river turned on its siren and faint sounds of the clamour reached me even above the roar of my motor. It seemed but a moment until the Statue of Liberty came into view. I turned westward, circled the Lady with the Torch and alighted safely on the parade ground on Governor's Island.

General Frederick Grant, commanding the Department of the East, was one of the first officers who came up to extend congratulations and to compliment me on the success of the undertaking. From that moment I had little chance for anything except the luncheons and dinners to which I was invited. First came the luncheon at the Astor House given by the *New York World*, and then the big banquet at the Hotel Astor, presided over by Mayor Gaynor and attended by many prominent men interested in aviation. The speeches were all highly laudatory, of course, and there were many predictions by the orators that the Hudson river would become a highway for aerial craft, as it had for steam craft when Fulton first steered the old *Clermont* from New York to Albany.

On the trip down from Albany I carried a letter

from the mayor of that city to Mayor Gaynor, and delivered it in less time than it would have taken the fastest mail train. My actual flying time was two hours, fifty-one minutes, the distance one hundred and fifty-two miles, and the average speed fifty-two miles an hour.

From Albany to Poughkeepsie is eighty-seven miles, and by making this in a continuous flight I had, incidentally, won the Scientific American trophy for the third time. It now became my personal property, and its formal presentation was made at the annual dinner of the Aero Club of America for that year.

NOTE BY AUGUSTUS POST

The newspapers made much of Mr. Curtiss' flight, drawing comparisons between the Hudson river course and the flight made by Blériot across the English channel, and the trip of Paulhan from London to Manchester, which he had just accomplished—a flight of about the same distance, for which he received fifty thousand dollars from the *London Daily Mail.*

The *New York Times* offered a large prize for a flight from New York to Philadelphia and return, immediately afterward, which Charles K. Hamilton won, and also offered a prize of twenty-five thousand dollars for a flight between New York and Chicago, which was never won. Mr. W. R. Hearst was also moved to offer fifty thousand dollars for a flight between New York and a point on the Pacific Coast, the offer standing open for one year. This flight was accomplished by Calbraith P. Rodgers, but was not concluded within the time limit.

There was, naturally, an outburst of editorial comment from newspapers all over the United States, not only long and scholarly leaders, but brief, snappy paragraphs that make the press of this country an interesting record of public feeling and sentiment on all extraordinary achievements. For instance, the *St. Louis Times* spoke of the passing of the new aerial menace over West Point where cadets were studying the history of military science along ancient lines, and the *Chicago Inter-Ocean* chuckled over how this latest achievement "would jar old Hendrik Hudson."

The *Newark News* declared that "the Indian canoe, the *Halfmoon*, the *Clermont* and the Curtiss biplane each represented a human achievement that marked an epoch," while the *Providence News* believed that "valuable as was astronomer Halley's naming of a comet, Mr. Curtiss has accomplished something of more practical value to the world" and the *York Gazette* compared the flight down the Hudson Valley by the aeroplane, to the conquest of the North Pole. There were other interesting points of view taken by the press, the *Birmingham News*, for instance, expressing the opinion that the *New York*

THE HUDSON FLIGHT
(A) Stop at Poughkeepsie. (B) Finish, at Governor's Island

THE EVOLUTION OF THE HYDRO

(A) The first hydro in the world—the "June Bug" on pontoons, Hammondsport, November 5, 1908. (B) Developing Hydro at San Diego—Curtiss and Ellyson in hydro of winter, 1911; dual control—either of two military aviators may steer. (C) Curtiss landing in hydro at Cedar Point, Ohio

World was extravagant, as "it had paid $10,-000.00 for Curtiss' ticket from Albany to New York, when it might have brought him down by train for $4.65." The *Battle Creek Enquirer* said that Mr. Curtiss ought to go into politics, for "a man who can soar as high, stay up as long, travel as far, light as safely, *all on wind,* would have the rest of them tied to the post." But the *Savannah News* intimated that nobody could blame Mr. Curtiss from flying away from the Albany Legislature at the rate of a mile a minute. The *Birmingham Age-Herald* declared that the way was paved for other and greater flights, even across the Atlantic ocean, and indeed, the ocean flight now seemed to the press a not far distant possibility. The *Rochester Chronicle-Democrat* argued that the bench and bar would now have an opportunity for the exercise of all their legal ability to settle the question "who owns the air?" But it was left to the *Houston Post* to break into poetry in the following outburst of local pride:

> "The wonder is that Curtiss did
> Not pass New York and onward whiz
> Southwest by south, half south, until
> He got where Houston, Texas, is."

But perhaps the most characteristic comments were those like that of the *New York Evening Mail:*

"In every newspaper that you picked up yesterday you read a thrilling account of the great achievement of Glenn H. Curtiss. The detailed description of his wonderful flight stirred every emotion in you. Chills ran up your spine and tears of joy came to your eyes as you read on and on of the courage of the man who propelled his airship at a speed of fifty-three miles an hour at a height of a thousand feet above the earth. He realised all of the time that a broken bolt or some little thing gone wrong might dash him to death."

It is of course quite impossible to give even a small proportion of the bright comments that were made by the newspapers not only of this country, but even by the foreign press. The *New York Times* sent a special train to follow the flight, on which I rode as the representative of the Aero Club of America. Here is my report in the *Times:*

"7:02 A. M.—Mr. Curtiss started from Van Rensselaer Island, Albany. Jacob L. Ten Eyck official starter for Aero Club of America.
7:03—Passed over the city limits of Albany.
7:20—New Baltimore.
7:26—Twenty-one miles. *The Times* special train caught up with aeroplane.
7:27—Milton Hook brick yards. Wind still. Aeroplane flying about 45 miles per hour. Passed lighthouse on west side of Hudson River.
7:32—Stockport. Twenty-four miles.
7:35—Hudson. Twenty-nine miles. Aeroplane flying high. Catskill Mountain houses could be seen in the distance.

THE HUDSON FLIGHT

Machine flying steady, water was calm, small ripples along the surface.

7:36—Thirty miles. *The Times* special train passed through tunnel parallel with 'plane.

7:40½—Tower 81, New York Central Railroad. Greensdale ferry.

7:41—Catskill on west shore of Hudson River. Flying high.

7:44½—Water trough in centre of track. Train equal with 'plane. Linlithgo Station.

7:46—Germantown steamer dock. Aeroplane flying well.

7:48—Passed old steamboat on west side of the river. Germantown Station. Aeroplane pitched when foot oil pump was used. Slight ripples on the water.

7:51—*The Times* special train running parallel with aeroplane.

7:53—Tivoli. Forty-four miles. Aeroplane 1,000 feet high. Wind slightly from the west.

7:58—Barrytown. Forty-nine miles. Aeroplane about 800 feet high, descending a little lower until about 400 feet high.

8:03—Kingston. Brick yards on west shore of river. Mr. Curtiss is flying very near *The Times* special train, within perhaps 100 yards.

8:04—Aeroplane turns toward west. Heads a little more into the wind and crosses to the west side of the river at high speed.

8:05—Private yacht dock on east side of river. Aeroplane flying high again.

8:06—Rhinecliff Ferry. Fifty-four miles. Aeroplane has been flying one hour and four minutes. Seems to be flying well.

8:08—Passing Tower 67, New York Central Railroad.

8:08½—*The Times* special train passed through tunnel. Mr. Curtiss goes back to west side of river, flying over icehouses.

8:11—Passed lighthouse in middle of river. The aeroplane seems to be rising and falling slowly on the varying currents of air. River is very wide at this point. There are large stone crushers on the west shore, and a large stone building of an institution on the bank of the river.

8:12—Staatsburg. Sixty miles.

8:16—Aeroplane now is passing over a large white house, some private residence on the west shore of the river. Aeroplane is flying past freight train on the West Shore Railroad.

8:18—Hyde Park Station. Sixty-four miles. *The Times* special train passing water trough in centre of railway track. Passing Insane Asylum at Poughkeepsie.

8:20—Passing upper portion of Poughkeepsie. 'Plane over river.

8:24—Passing Poughkeepsie Bridge. Aeroplane about 200 feet above it.

8:25½—*The Times* special train goes through Poughkeepsie Station.

8:30—*The Times* special train arrives at Gill's Mill Dock, opposite landing place of Mr. Curtiss. Aeroplane landed according to Mr. Curtiss's watch on his machine at 8:26. I left special train and went to the field where Mr. Curtiss had landed, arriving a few minutes later. The tanks of the machine were filled with eight gallons of gasoline and one gallon and a half of oil. The machine was examined carefully and found to be in good order, one wire being stayed to prevent vibration. George Collingwood took *The Times* special train party to New Hamburg Station.

9:26—Mr. Curtiss started for New York from field on property of Mr. Gill.

9:31—Camelot.

10:02—West Point. Aeroplane passed over Constitution Island at an altitude of about 400 feet above the land.

10:06—Manitou.

10:14—Peekskill.

10:15—Ossining. Aeroplane flying on west side of the river.

THE HUDSON FLIGHT

10:25—Dobbs Ferry.
10:30—Yonkers. Aeroplane flying about level with top of Palisades.
10:35—Landed 214th Street.—Inwood. After passing down river to Dyckman Street and returning to Spuyten Duyvil and passing over drawbridge the aeroplane landed upon the property of the Isham estate.
11:42—Mr. Curtiss left his landing place, flying again over the drawbridge, out over the Hudson River, turned south.
12:00 M.—Passed New York City and landed at Governor's Island at noon.

"Mr. Curtiss also entered for the Scientific American trophy and the first flight from Albany to the landing place at Poughkeepsie, the exact distance of which is to be determined later, will count as a record for this event, and if not exceeded in the year will stand as Mr. Curtiss's trial for this trophy.

"The figures as finally corrected show that Mr. Curtiss was in the air on the first leg of his flight from Albany to the Gill farm near Poughkeepsie 1 hour and 24 minutes; from the Gill farm to the Isham estate at 214th Street 1 hour and nine minutes, and from 214th Street to Governor's Island 18 minutes, making a total flying time for the 150 miles of 2 hours and 51 minutes.

"Figured on the basis of 150 miles for the entire flight, Mr. Curtiss is shown to have maintained an average speed of 52.63 miles per hour."—A. P.

CHAPTER IV

THE BEGINNING OF THE HYDROAEROPLANE

THE Albany Flight was a great stimulus to aeronautics in this country. Prizes were at once offered in several different places by several different newspapers, and a great many cities wanted to have public flights made and particularly wanted flights to be made over water.

At Atlantic City I flew over the ocean, making a record for fifty miles over water on a measured course. It was here at the same time that Walter Brookins made a world's altitude record of over six thousand feet in a standard Wright machine. Later I flew from Cleveland to Cedar Point, near Sandusky, Ohio, a distance of sixty miles over the waters of Lake Erie, and returned next day in a rain storm.

After making flights in Pittsburgh, Pa., I thought that a successful meet could be held in New York City, so I arranged to have all of our forces gathered together at Sheepshead Bay race track, near Brighton Beach, N. Y., and during the week of August 26, 1910, we had an aeroplane meet at which Messrs. J. C. Mars, Charles F. Willard, Eugene B. Ely, J. A. D. McCurdy, and Augustus Post made flights and this meet was so successful that it was continued for a second week. Mr. Ely flew to Brighton Beach and took dinner and then flew back. Mr. Mars flew out over the Lower Bay and we had all five of the machines in the air at one time on several occasions—a record for New York at that time. It was here that Mr. Post made a Bronco Busting Flight over the hurdles at the Sheepshead Bay track, landing safely after putting his machine through all manner of thrilling manœuvres.

The Harvard Aeronautical Society had arranged a meet at Boston, Mass., which followed directly after this one, and Claude Grahame-White, the famous English aviator, who was later to win the Gordon Bennett cup at Belmont Park, came over from England, bringing his fast Blériot monoplane with him. A special race was arranged between Mr. White in his Blériot and my racing biplane. The meet was a great success, and but a very small margin separated Mr. White's Blériot and my machine when we tried out our best speeds.

Then came a meet at Chicago,[1] after which it

[1] NOTE BY AUGUSTUS POST

While flying in the Chicago meet we had four machines in the air at once. I was a novice at flying then but entered the air while the other fellows were flying around.

Circling the track I was just passing the grand stand when Willard swooped down in front of me having passed right over my head.

was arranged that three machines should start to fly from Chicago to New York for the *New York Times'* prize of $25,000. A team was made up and Mr. Ely was chosen to make the attempt to fly to New York. This was a very ambitious undertaking for this period in the history of aviation in America, for the longest flight that up to this time had been made in this country was between New York and Philadelphia, one hundred and eighty miles; while the distance between Chicago and New York was fully one thousand miles and landings were very difficult to accomplish in the broken country along the way. Mr. Ely made a good attempt, but there was not sufficient time to complete the trip as flights had already been arranged at Cleveland, Ohio, and in order to go there, this attempt was given up.

The Gordon Bennett Aviation Cup race was the

I clung on to the steering post and held the wheel as firmly as I could while to my great consternation the machine rocked and swayed fearfully in the back draft from Willard's propeller. He kept doing the Dutch Roll and the Coney Island Dip right in front of me, which made it all the worse, as the wash of the propeller wake would strike above and below my machine as he pitched up and down in front of me. I stood it as best I could, hardly daring to breathe but holding my course and balancing with all my might, until Willard turned off, and then after a bit I made a good landing. When Willard came down he rushed up to me and grabbed me by the hand and said, "Oh, Post! will you ever forgive me for that? I ought to have known better than to back-wash you but you know I thought you were Ely, and I wanted to scare him!"—A. P.

BEGINNING THE HYDRO 115

next thing to arouse the interest of patriotic Americans and the Aero Club of America had been busy with arrangements for a big meet to be held at Belmont Park, near New York. This was the largest undertaking that the club had up to this time attempted and they taxed every possible resource, with the splendid result of securing all the foremost fliers of Europe, as well as of America, to participate.

I had built a machine for the trials which I thought would be very fast and had constructed it as a type of monoplane in order to cut down the head resistance to the very least possible point. America was represented by Anthony Drexel, Jr., in a Blériot; by the Wright Brothers, who had constructed a racing machine by putting a powerful motor in a small machine which was about one-half the size of their regular model, and by Mr. Charles K. Hamilton, who flew a Curtiss type machine, but with a large power motor of another make. Mr. Grahame-White won the race in his Blériot, although Mr. Alfred Leblanc, representing France, made remarkable time, but on the last lap ran into a telegraph pole on one of the turns and smashed his machine and had a most miraculous escape from being killed.

I did not try out my monoplane, although my regular type was the speediest standard biplane at the meet and was very well handled by Ely,

Mars, Willard, and McCurdy who flew in the contests. I had given up public flying in contests at this time.

A new line of thought—or to express it more accurately, the following out of a very old one—was taking my interest and a great part of my time. The experiments I had in mind involved the problem of flying from the water and alighting on the water.

The season of 1910 was now far advanced and it was time to make plans for the winter. Flying meets were to be held at Los Angeles again, and also at San Francisco, and California seemed the best place to go, for the weather there would be most favourable not only for winter flying, but also for carrying on the experiments which I had in mind. Meantime, when it seemed as if all the paths were open to the aeroplane over the land, and it was only a question of development, not of pioneering, it was suggested to me by the *New York World* to launch an aeroplane from the deck of a ship at sea and have it fly back to shore carrying messages.

The Hamburg American Steamship Company offered their ocean liner *Pennsylvania* for this test, and I sent a standard Curtiss biplane to be operated by J. A. D. McCurdy. The ship was fitted with a large platform, erected on the stern, a platform sloping downward, and wide enough to allow an aeroplane set up on it to run down

BEGINNING THE HYDRO 117

so that it could gather headway for its flight. The plan was to take McCurdy and the aeroplane fifty miles out to sea on the outward voyage from New York, and then launch them from the platform.

A mishap at the last moment upset all the well-laid plans. In trying out the motor just as the *Pennsylvania* was about to leave her dock at Hoboken, an oil can, carelessly left on one of the planes by a mechanic, was knocked off and fell into the whirling propeller. The result was a broken propeller, and as the ship could not delay its sailing long enough for us to get another, the attempt was abandoned.

In the meantime, however, the Navy became interested in the sea experiments and offered the armoured cruiser *Birmingham,* then at Hampton Roads, to be fitted up with a similar platform for launching an aeroplane. This was accepted and Eugene Ely, who was flying in a meet at Baltimore and already in the vicinity of Norfolk, took his Curtiss biplane over to the *Birmingham* for the test, fired with enthusiasm by McCurdy's attempt. On November 14 the *Birmingham,* equipped with a platform for starting the aeroplane, awaited good weather for the flight. The good weather did not come and after waiting impatiently on board for some time, Ely determined to risk a start, even though there was a strong wind coming off shore carrying a heavy mist that made it almost impossible to see more than half a mile. The ship was at anchor, but starting up his motor he flew off with the greatest ease, slightly touching the water with the wheels of his machine, but quickly rising and flying straight to shore, where he landed without difficulty.

This flight attracted world-wide attention, especially among the officers of the navies of the world. It was the first demonstration of the claims of the aeronautical enthusiasts of the navy that an aeroplane could be made that would be adaptable to the uses of the service, and it appeared to substantiate some of the things claimed for it.

When I found that business would bring me to California during the winter, and probably would keep me there for several months, I decided to grasp the opportunity to do the development work I had long wanted to do, and at the same time to request the honour of instructing representative officers of the Army and Navy in the operation of the aeroplane. I believed the time had arrived when the Government would be interested in any phase of aviation that promised to increase the usefulness of the aeroplane for military service.

So, on November 29, 1910, I sent letters to both Secretary Dickinson of the War Department and to Secretary Meyer of the Navy Department, inviting them to send one or more officers of their respective departments to Southern California,

BEGINNING THE HYDRO

where I would undertake to instruct them in aviation. I made no conditions. I asked for and received no remuneration whatsoever for this service. I consider it an honour to be able to tender my services in this connection. Other governments had already organised their aeronautical military branches and instructed men to fly, and it seemed to me that our own Government would do likewise were the opportunity afforded the officers to familiarise themselves with the aeroplane.

The invitations to the War and Navy Departments were written just prior to my departure for the Pacific Coast, and three weeks later I was notified that the Secretary of the Navy had accepted, and that they would detail officers for instruction.

It began to look, even to the doubters, as if an aeroplane could be made adaptable to the uses of the Navy, as the aeronautic enthusiasts of the service had claimed. The experiment begun would have to be completed, however, by flying from shore to the vessel, and for this opportunity we were eager. The chance came when we were all at San Francisco and another *Pennsylvania*, this time the big armoured cruiser, was in the bay. Rear Admiral Thomas, and Captain Pond, in command of the *Pennsylvania*, readily consented to assist in these further experiments. The *Pennsylvania* went to Mare Island to be outfitted, Ely and I going there to tell the Navy officials at the station just what would be required for such a hazardous test.

The platform was like that built on the *Birmingham*, but in the case of a flight to, instead of from, a ship the serious problem is to land the aeroplane on the deck and to stop it quickly before it runs into the masts of the ship, or other obstructions. The platform was built over the quarterdeck, about one hundred and twenty-five feet long by thirty feet wide, with a slope toward the stern of some twelve feet. Across this runway we stretched ropes every few feet with a sand bag on each end. These ropes were raised high enough so they could catch in grab-hooks which we placed under the main centrepiece of the aeroplane, so that catching in the ropes the heavy sand bags attached would drag until they brought the machine to a stop.

To protect the aviator and to catch him in case he should be pitched out of his seat in landing, heavy awnings were stretched on either side of the runway and at the upper end of it.

When all arrangements had been completed, and only favourable weather was needed to carry out the experiment, I was obliged to leave for San Diego, and, therefore, was unable to witness the flight. I regarded the thing as most difficult of accomplishment. Of course, I had every faith in Ely as an aviator, and knew that he would arrive

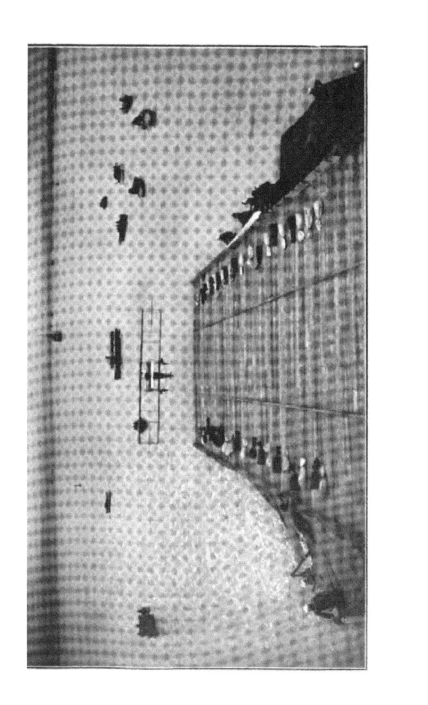

TWO FAMOUS MILITARY TEST FLIGHTS
(A) Curtiss and hydro hoisted on U. S. S. "Pennsylvania," at San Diego.
(B) Ely leaving "Pennsylvania." San Francisco harbor

BEGINNING THE HYDRO 121

at the ship without trouble, but I must confess that I had misgivings about his being able to come down on a platform but four feet wider than the width of the planes of the aeroplane, and to bring it to a stop within the hundred feet available for the run.

Ely rose from the Presidio parade grounds, flew out over the bay, hovered above the ship for an instant, and then swooped down, cutting off his power and running lightly up the platform, when the drag of the sand bags brought him to a stop exactly in the centre, probably one of the greatest feats in accurate landing ever performed by an aviator. As I have said, the platform was only four feet wider than the planes of the Curtiss biplane that Ely used, yet the photograph taken from the fighting top of the ship shows the machine touching the platform squarely in the centre. When one stops to think that the aeroplane was travelling about forty miles an hour when it touched the deck and was brought to a stop within a hundred feet, the remarkable precision of the aviator will be appreciated.

Not only was there not the least mishap to himself or to the machine in landing, but as soon as he had received a few of the many excited congratulations awaiting him, he started off again and flew back the ten miles to the camp of the 30th Infantry on the Aviation Field, where wild cheers greeted the man and the machine that had for the first time linked the Army and the Navy. For this is what, in the wars of the future, or even in the preservation of the future's peace, the aeroplane is certainly going to do, joining as nothing else can the two branches of the service.

I don't think there has ever been so remarkable a landing made with an aeroplane as Ely's, and probably never so much store put by the mere act of coming down in the right place. A few feet either way, a sudden puff of wind to lift the aeroplane when it should descend, or any one of a dozen other things, might have spelled disaster for the whole undertaking, deprived the daring aviator of a well earned success, and the world of a remarkable spectacular demonstration of practical aviation.

On the day of the test I was in San Diego and awaited news from San Francisco with a good deal of impatience. When at last the Associated Press bulletin announced that Ely had landed without mishap I first felt a great relief that there had been no accident to mar the success of the thing, and then a sense of elation that we had taken another long step in the advancement of aviation.

Early in January I went to Southern California to establish an experimental station, and at the same time to instruct the officers of the Army and Navy whom I had invited the War and Navy Departments to assign for that purpose. A part

BEGINNING THE HYDRO   123

of our experiments were along the line of a new "amphibious" machine that had been on my mind ever since my first experiments in Hammondsport.

I believed that with the proper equipment for floating and attaining a high speed on the water, an aeroplane could be made to rise as easily as it could from the land.[1] I had carried these experiments just far enough in Hammondsport to convince me that the thing was feasible, when I was obliged to discontinue them to take up other business. I knew it would be safer to land on the water than on land with the proper appliances, and that it would be easier to find a suitable landing place on water, for the reason that it always

---

[1] NOTE BY AUGUSTUS POST

An interesting story is told of how the hydroaeroplane came to be invented.

During the period when he was planning a new series of experiments, Mr. Curtiss, accompanied by Mrs. Curtiss, attended a New York theatre in which there was being presented a play much talked about just then. The curtain went up on the first act, and the noted aviator was apparently enjoying the show when, just as the scene was developing one of its most interesting climaxes, he turned to Mrs. Curtiss and said: "I've got it." On the theatre program he had sketched what ultimately became the design of the hydroaeroplane.

This is like a time when Mr. Curtiss was standing one day by the side of one of his motorcycles talking with a customer. He kept turning one of the grips of the handle-bar with his fingers while talking and after finishing the conversation went into his office and developed the idea of a handle-control which had come to him while apparently absorbed in conversation.—A. P.

affords an open space, while it is often difficult to pick a landing place on the land. So, when I made preparations for my flight from Albany to New York City, I fitted pontoons beneath the chassis of my machine and a hydro-surface under the front wheel. I wanted to be prepared for alighting on the water should anything go amiss. As a matter of fact, the river course was the only feasible one for this flight, as there were mountains and hills for almost the entire distance.

It was while on that trip that I decided to build an aeroplane that would be available for starting or landing on the water. I don't know that I had the idea of its military value when I first planned it; but it came to me later that such a machine would be of great service should the Navy adopt the aeroplane as a part of its equipment. I thought the next step from pontoons, to float an aeroplane safely on the water, would be a permanent boat so shaped that it could get up speed enough so the whole machine could rise clear of the water and fly in the air.

It was important to find a location where it would be possible to work along the lines I had mapped out—a place where I might be free from the pressing calls of business and the hampering influence of uncertain climatic conditions. In short I wanted a place with the best climate to be found in this country, with a field large enough and level enough for practice land flights by be-

BEGINNING THE HYDRO

ginners, and with a convenient body of smooth water for experiments with a machine that would start from or land upon water.

Above all, I wanted a place not easy of access to the curious crowds that gather wherever there is anything novel to be attempted; for a flying machine never loses its attraction to the curious. Mankind has been looking for it ever since the beginning of the world, and now that it is actually here he can't get away from it, once it is in sight. A machine that has actually carried a man through the air takes on a sort of individuality all its own that acts as a magnet for the inquiring mind. Once people have really seen an aeroplane fly, they want to know what makes it fly and to come into personal contact with the machine and the man who operates it.

San Diego was brought to my attention as affording every advantage for experimental work in aviation. A study of the weather bureau records here showed a minimum of wind and a maximum of sunshine the year round. I visited that city in January, 1911, and after a thorough inspection of the grounds offered as an aviation field, decided to make that city the headquarters for the winter and to carry on the experimental and instructional work there.

North Island, lying in San Diego Bay, a mile across from the city, was turned over to me by its owners, the Spreckels Company. It is a flat, sandy island, about four miles long and two miles wide, with a number of good fields for land flights. The beaches on both the ocean and bay sides are good, affording level stretches for starting or landing an aeroplane. Besides, the beaches were necessary to the water experiments I wished to make. North Island is uninhabited except by hundreds of jack rabbits, cottontails, snipe, and quail. It joins Coronado Island by a narrow sand spit on the south side, which is often washed by the high tides. Otherwise the two islands are separated by a strip of shallow water a mile long and a couple of hundred yards wide, called Spanish Bight. Thus the island on which we were to do our experimenting and training was accessible only by boat and it was a comparatively easy matter to exclude the curious visitor whenever we desired to do so. There was no particular reason for excluding the public other than the desire to work unhampered by crowds, which is always a distracting influence.

In the meantime Lieutenant Theodore G. Ellyson of the submarine service, then stationed at Newport News, Virginia, had been detailed by the Navy Department to report to me in California for instruction in aviation. He had joined me in Los Angeles, where, though there are all the climatic requirements, and good fields for practice flights, the ideal body of smooth water for experiments on that element was lacking. The

BEGINNING THE HYDRO

War Department responded later, instructing General Bliss, commanding the Department of California at San Francisco, to detail as many officers as could be spared to go to San Diego for instruction in the art of flying.

There was much eagerness among the officers of the Department of California and I was informed that some thirty applications were made for the detail. Lieutenant (now Captain) Paul W. Beck, of the Signal Corps, located at the Presidio, San Francisco, and Lieutenant John C. Walker, Jr., of the 8th Infantry, Monterey, Cal., were named at once, and later Lieutenant C. E. M. Kelly, 30th Infantry, San Francisco, was added to the Army's representation. This made a list of four officers, three from the Army and one from the Navy, and with these I began work. In February, however, the Navy Department designated Ensign Charles Pousland of the destroyer *Preble*, at San Diego, to join Lieutenant Ellyson as a Navy pupil in aviation.

There are a dozen good landing or starting fields on North Island, but we chose the one on the south side, which gave us easy access to the smooth shallow water of Spanish Bight. A field was cleared of weeds and sagebrush, half a mile long by three or four hundred yards wide. Sheds to house the machines were built by the Aero Club of San Diego, and landings put in for the small boats that carried us to and from the city.

The Spreckels Company gave us every assistance in fitting the place up, and the people of San Diego, anxious to make the island the permanent home of an aviation experimental station and school, were prompt to lend a hand and to impress upon us the climatic advantages of their city.

I have asked Lieutenant Ellyson to write his own story of the work on North Island, and it is to be found in another part of this book.

CHAPTER V

DEVELOPING THE HYDROAEROPLANE AT SAN DIEGO—
THE HYDRO OF THE SUMMER OF 1912

JANUARY had nearly passed before the first machine was ready. Although this proved unsuccessful, I was not discouraged and learned a good deal about what sort of a float was necessary to support the aeroplane and how it acted when under way over the water. Nearly every day for over two weeks we dragged the machine down to the edge of the water, launched it on the smooth surface of San Diego Bay, and drew it out again after testing out some new arrangement of floats and surfaces. We kept it in a hangar, or shed, on the beach, and there we would sit and study and change and plan how to improve the float.

We were in the water almost all day long; no thought was given to wet clothing and cold feet. We virtually lived in our bathing suits. The warm climate aided us, but there were some chilly days. Discomfort and failure did not deter the Army and Navy officers, who watched and worked like beavers, half in and half out of the water.

On the 26th of January the first success came. That day the aeroplane first rose from the water and succeeded in alighting gently and without accident after the flight. A page was added to aviation history, which extended its domain and opened the lakes, rivers, and seas to the hitherto land-locked flying machine. It was no more a land bird, but a water fowl as well.

The machine was crude, and there remained many things to be improved, but the principle was correct. We kept adjusting the equipment, adding things and taking them off again to make some improvement; perhaps the float was too heavy, or leaked, or the spray would fly up and chips would be knocked out of the whirling propeller, which the drops of water would strike like shot out of a gun. The least projection on the floats would send up spray while travelling at such high speed as was made through the water. The balance of the machine was as troublesome as anything, because the push of the propeller would give it a tendency to dive if the floats were not properly adjusted.

When we brought the machine out on the 26th day of January I felt that we ought to get some results. There were no crowds of people present and there was no announcement of what was about to happen. I had not expected to make a flight, but climbed into the aviator's seat with a feeling that the machine would surely rise into the air when I wished, but that I would only try it on the water to see how the new float acted.

DEVELOPING THE HYDRO 131

Lieutenant Ellyson spun the propeller and I turned the machine into the wind. It ploughed through the water deeply at first, but gathered speed and rose higher and higher in the water and skipped more and more lightly until the float barely skimmed the surface of the bay. So intent was I in watching the water that I did not notice that I was approaching the shore and to avoid running aground I tilted the horizontal control and the machine seemed to leap into the air like a frightened gull. So suddenly did it rise that it quite took me by surprise.

But I kept the machine up for perhaps half a mile, then turned and dropped lightly down on the water, turned around and headed back to the starting point. The effect of that first flight on the men who had worked, waited, and watched for it was magical. They ran up and down the beach, throwing their hats up into the air and shouting in their enthusiasm.

I now headed about into the bay, in the direction of San Diego, and rose up into the air again even more easily than the first time. I flew for half a mile and turned twice to see how the machine would act in the air with the clumsy-looking float below it. The naval repair ship *Iris* caught sight of me as I went flying by and sent its siren blast far out over the water, and all the other craft blew their whistles, until it seemed as if all San Diego knew of the achievement. Satisfied that it was all right, I landed within a few yards of the shore, near the hangar.

We made flights nearly every day after this, taking the Army and Navy officers as passengers. I found the machine well adapted for passenger work and it became very popular. While experimenting we kept changing things from day to day, adding and taking off, lightening the machine, or adding more surface. We tried putting on an extra surface, making a triplane, and got remarkable lifting power. We changed the floats and finally made one long, flat-bottomed, scow-shaped float, twelve feet long, two feet wide, and twelve inches deep. It was made of wood, the bow being curved upward the full width of the boat and at the stern being curved downward in a similar manner. This single float was placed under the aeroplane so that the weight was slightly to the rear of the centre of the float, causing it to slant upward, giving it the necessary angle for hydroplaning on the surface of the water.

I will confess that I got more pleasure out of flying the new machine over water than I ever got flying over land, and the danger, too, was greatly lessened.

I then decided upon a test which I had been informed the Navy regarded as very important. In fact, I had been told that the Secretary of the Navy regarded the adaptability of the aeroplane to navy uses as depending very largely on its

DEVELOPING THE HYDRO 133

ability to alight on the water and be hoisted aboard a warship. With the hydroaeroplane I had developed, I had no doubts about being able to do this, without any platform or preparation on board the vessel.

So, on February 17, at San Diego, I sent word over to Captain Charles F. Pond, commanding the armoured cruiser *Pennsylvania,* then in the harbour, that I would be pleased to fly over and be hoisted aboard whenever it was convenient to him. He replied immediately, "come on over." The *Pennsylvania* is the ship that Ely landed on at San Francisco in his memorable flight, and it was Captain Pond who at that time gave over his ship and lent every assistance in his power to make the experiment the success it was. He lent his aid to this second experiment as willingly as he did to the first.

There were no special arrangements necessary for this test. All that would be needed to get the aeroplane and its operator on board would be to use one of the big hoisting cranes, just as they are used for handling the ship's launches.

The hydroaeroplane was launched on Spanish Bight, and in five minutes I was on the way. The machine skimmed over the water for a hundred yards and then rose into the air. In two or three minutes I was alongside the cruiser, just off the starboard quarter. There was a strong tide running and when I shut off the propeller the aeroplane drifted until a rope thrown from the ship was made fast to one of the planes by Lieutenant Ellyson of the Navy. It was drawn in close to the side of the ship, where a boat crane was lowered and I hooked it in a wire sling attached to the top of the planes. I then climbed up on top of the aeroplane and slipped my leg through the big hook of the crane, not caring to trust too much weight to the untested sling.

In five minutes from the time I landed on the water alongside the ship, the hydroaeroplane reposed easily on the superstructure deck of the big cruiser, just forward of the boat crane. It had been the easiest sort of work to land it there, and thus one more of the problems that stood in the way of a successful naval aeroplane was overcome.

The rest of the experiment was performed with equal promptness and ease. After a stay of ten minutes on the cruiser, the aeroplane was dropped overboard by the big boat crane, the propeller was cranked by one of the military pupils in aviation, and I got under way for the return trip to the island. Two minutes later I brought the hydroaeroplane to a stop a few yards away from the hangar on the beach. The entire time taken from the moment I left North Island for the cruiser to the moment I landed on the water at the hangar on my return was less than half an hour, and yet within this brief space had been written one of the most interesting chapters in the history of naval aviation.

I regard this experiment as one of the most interesting, from my idea of a military experiment, that had been attempted up to that time, for the reason that no special equipment was needed on board the ship. Obviously the objections to the landing of an aeroplane on deck from a flight had to be overcome, and this could be done with a machine that could land on the water and be picked up. For a flight from the ship, all that was necessary was to drop it over the side and watch it rise from the water into the air. Such a machine could be "knocked down" and stored in a very small space when not in use; and when wanted for a flight, it could be brought out and set up in a short time on deck.

An aeroplane sent from a scout ship on a scouting flight must, to be efficient, be able to carry a passenger, especially if it be sent for any purpose other than as a messenger, where speed would be the first consideration. But if sent to seek information as to an enemy's position, to take observations and make maps of the surrounding country, or with any of a dozen other objects in view where a trained observer would be necessary, it seems to me it should be equipped to carry at least two, and possibly three, persons—the aviator and two passengers. There were many machines capable of carrying one or more passengers on land flights, so I set about equipping one to carry passengers on water flights.

This I first succeeded in doing on February 23, when I took up Lieutenant T. G. Ellyson of the Navy, in the hydroaeroplane. We rose from the water without difficulty, flew over San Diego Bay and returning, alighted on the water with perfect ease.

This was all very well and good where a flight was to be made from the water and back to the water; but I believed we should go further and provide a machine that would be able to go from one to the other—from water to land and land back to water—before it could be said that all the difficulties of making the aeroplane adaptable to both Army and Navy uses had been overcome. This was of comparatively easy accomplishment, and on Sunday, February 26, I made the first flight from water to land and from land back to water. Starting from North Island, on the waters of Spanish Bight, I flew out over the ocean and down the beach to a point near Coronado Hotel, where I came down on the smooth sand of the beach. Returning, the machine started from the beach and came back to the water on Spanish Bight whence I had started.

With these achievements it seems to me the aeroplane has reached the point of utility for military purposes—either for the Army or Navy. It now seems possible to use it to establish communication between the Navy and Army, when there are no other means of communication. That is, a warship could launch an aeroplane that can fly over sea and land and come to earth on whichever element affords the best landing. Having fulfilled its mission on shore it could start from the land, and, returning to the home ship, land at its side and be picked up, as I was picked up and hoisted aboard the *Pennsylvania* at San Diego.

Here let me call attention to the splendid field that California offers for the development of aviation, with its climate, permitting aviation to be pursued all the year, and its large winter tourist population with wealth and leisure to devote to furthering the art of flight. In California even the legislature recognises the increasing popularity of flying, and it has given careful attention to the formation of laws to protect the aeroplane and the aviator.

There remained one thing further to accomplish complete success with the hydroaeroplane, and that was to devise a method of successfully launching the machine from a ship without touching the water and without resorting to any cumbersome platform or any other launching apparatus that would interfere with the ship's ordinary working. To accomplish this would solve the principal obstacle that stood in the way of using the hydroaeroplane at sea.

Lieutenant Theodore G. Ellyson, of the United States Navy, had been working out a plan for doing this and it was not until September, 1911, that the experiment was finally completed at Hammondsport, where operations were continued after breaking up the camp at San Diego, late in the spring.

A platform sixteen feet high was erected on the shore of Lake Keuka and a wire cable two hundred and fifty feet long was stretched from the platform to a spile under water out in the lake. The hydroaeroplane was set on this wire cable near the platform on which the men stood to start the propeller. A groove was made along the bottom of the boat in which the cable fitted loosely, to guide it as it slid down, until sufficient headway was obtained to enable the wings of the aeroplane to support the weight of the machine. A trial of this method of launching was entirely successful. The machine started down the cable gathering headway and we all watched it gracefully rise into the air and fly out over the lake. This launching from a wire is the last step in the development of handling the aeroplane and it is hardly possible to foresee all the many important applications which will be made in the future of this type of machine, since a cable can be easily stretched from the bow of any vessel, which can then steam into the wind, easily enabling an aeroplane to be launched in almost any weather, while

DEVELOPING THE HYDRO 139

it can without difficulty land under the lea of the vessel and be hoisted on board again.

As the wireless has almost revolutionised ocean navigation by furnishing a means of constant communication between steamers, perhaps the hydroaeroplane will be able to bring passengers back to shore or take them from shore to a ship on the high sea, or enable visits to be made between ships that pass on the ocean. Great, powerful hydroaeroplanes may be able to cross the ocean itself at high speed, and they will no doubt add greatly to the safety of ocean travel, as well as furnish the Navy with an arm of destruction much more far-reaching than its most effective guns or torpedoes.

Frank Coffyn in May, 1912, took a belated passenger from the Battery, New York City, out to a steamer as it was steaming out of the lower bay and landed him safely aboard—a hint of future possibilities.

We had a curious opportunity to prove how the hydroaeroplane can be an arm of preservation as well as destruction, when at the Chicago meet of 1911. Simon, dashing over the lake, dropped in his machine. Hugh Robinson had been putting a hydroaeroplane through its evolutions, to the great interest of the crowd, who evidently thought it a sort of freak machine, but when Simon fell Robinson was after him instantly, and for the first time in the history of the world, a man flew through the air from dry land, alighted on the water beside a man in distress, and before anything else could get there, invited him to fly back to shore with him. As there were boats close at hand, the offer was not needed, but the value of the land-air-water machine had been proved, for it had left its hangar and flown a mile from shore in a little more than a minute.

The hydroaeroplane can already fly sixty miles an hour, skim the water at fifty miles, and run over the earth at thirty-five miles. Driven over the surface of the water the new machine can pass the fastest motor boat ever built and will respond to its rudder more quickly than any water craft afloat. Its appeal will be as strong to the aquatic as to the aerial enthusiast.

Flying an aeroplane is thrilling sport, but flying a hydroaeroplane is something to arouse the jaded senses of the most blasé. It fascinates, exhilarates, vivifies. It is like a yacht with horizontal sails that support it on the breezes. To see it skim the water like a swooping gull and then rise into the air, circle and soar to great heights, and finally drop gracefully down upon the water again, furnishes a thrill and inspires a wonder that does not come with any other sport on earth.

The hydroaeroplane is safer than the ordinary aeroplane, and for this reason is bound to become the most popular of aerial craft. The begin-

DEVELOPING THE HYDRO 141

ner can take it out on his neighboring lake or river, or even the great bays, and skim it over the water until he is sure of himself and sure that he can control it in the air. He can fly it six feet above the water for any distance, with the feeling that even if something should happen to cause a fall, he will not be dashed to pieces. The worst he will get is a cold bath.

The hydroaeroplane may compete with motor boats as a water craft, or in the air with the fastest aeroplane. It can start from the land on its wheels, but launch itself on the water where there is lack of room for rising from the land.

Its double qualities as a water and air craft make possible flights that could not be attempted with the aeroplane.

At Cedar Point, Ohio, I had to fly the new machine when a strong gale was blowing across Lake Erie, kicking up a heavy surf. However, I determined to make the attempt under what were extremely trying conditions, and so started it on the beach and under the power of the aerial propeller, launched it through a heavy surf.

Beyond the surf I found very rough water, but turning the machine into the wind, I arose from the water without the least difficulty, and circled and soared over the lake for fifteen minutes. I landed without trouble on the choppy water a few hundred yards off shore, and after guiding the hydroaeroplane up and down the beach for the inspection of the great crowd, made a second flight of ten minutes' duration, and landed safely upon the sandy beach. That was the hardest test I have ever given the hydroaeroplane, and I think a very severe one. I am satisfied that it can be used in more than ordinarily rough water, if it is properly handled.

There is no question that in this particular line of aeronautics, America is now leading the world; but the hydroaeroplane contests recently held at Monte Carlo and the experiments made in France by the Voisin Brothers' "Canard," which was erroneously hailed by the French press as being the first occasion when a machine had risen from the water with two men, show that the French are not far behind us.

Other experiments have been made in Europe by Fabre, who was the first to achieve any degree of success in this line, and by the Dufaux Brothers on the Lake of Geneva, to say nothing of the flights made by Herbster, the old Farman pilot, on an Astra-Wright at Lucerne, and if the American aeronautic industry does not awaken to the immediate possibilities along this line, it will once more be overtaken by Europeans.

There are thousands of men throughout the country who would gladly take up a new mechanical sport as a successor to motor boating and motoring if they felt they could do so with a reasonable degree of safety to themselves, and adequate assurance that the life of their machine would be commensurate to the price paid for it.

Followers of the sport of motor boating, which has made thousands of converts during the past few years, are already turning to the hydroplane, which skims over the water at much greater speed and less power. The next step will be the hydroaeroplane, which can skim over the water in exactly the same way and has the further enormous advantage of rising into the air whenever the driver so desires. The sport should develop rapidly next summer and be in full swing in a few years. Several improvements of detail will have to be made. Ways of housing the craft—of stopping the engine—of muffling the roar of the motor, will be devised; while more comfort for the pilot and passengers will be arranged.

If a cross-country flight is too dangerous to attempt because of the rough character of the land, the hydroaeroplane can follow a river course with perfect safety. Or, if there is no water course and the country is level, it can take the land course with equal safety.

In short, it matters little whether an aerial course takes one over land or water, the hydroaeroplane is the safest machine for flight. With the "Triad," as we called the machine from its triple field—air, land, and water—the Great

Lakes offer no impassable obstacle to a long flight, and it is within the vision of him who watches the trend of things, that an over-sea flight is not far in the future.

Note by Augustus Post

THE "FLYING BOAT"

At San Diego, on Jan. 10, 1912, a new type of Curtiss hydroaeroplane, or "flying boat," was given its first trial on the bay. It had been designed and constructed under strict secrecy at Hammondsport. The public knew nothing as to the details of this craft until it was taken out on the bay in order to test its balance and speed on the water.

This craft, which was equipped to carry a passenger, was driven by a sixty horse-power motor. In contact with the water, it went at over fifty miles an hour; and lifted off the water, it travelled at more than sixty miles an hour in the air. It differs in many respects from the hydroaeroplane now in use by the United States Navy officers who, by the way, were present and witnessed the test. There were two propellers instead of one and these were driven by clutch and chain transmission. They were really "tractors," being in front of the planes; the motor had a new automatic starter, and there was also a fuel gauge and

DEVELOPING THE HYDRO 145

bilge pump. The transmission has since been changed to direct drive.

The boat, or hydro equipment, contained a bulkhead fore and aft, was twenty feet long, with an upward slope in front and a downward slope in the rear. The hydro equipment, which was more like a boat than anything yet designed, was able to withstand any wind or wave that a motor boat of similar size could weather. The aviator sat comfortably in the hull with the engine not behind him, but for'ard in the hull in this model.

THE "FLYING FISH"

A "No. 2 flying boat," just built by Mr. Curtiss, and successfully tested on Lake Keuka, Hammondsport, in July, 1912, is the "last word" in aviation so far. An illustration in this book, made from photographs taken in mid-July, 1912, shows fully the bullet-shape of the "flying fish."

It is a real *boat*, built with a fish-shaped body containing two comfortable seats for the pilot and passenger or observer, either of whom can operate the machine by a system of dual control, making it also available for teaching the art of flying.

All the controls are fastened to the rear of the boat's hull, which makes them very rigid and strong, while the boat itself, made in stream-line form, offers the least possible resistance to the air, even less than that offered by the landing gear upon a standard land machine. Above the boat are mounted the wings and aeroplane surface. In the centre of this standard biplane construction is situated the eighty horse-power motor with its propeller in the rear, thus returning to the original practice, as in the standard Curtiss machines, of having a single propeller attached direct to the motor, thus doing away with all chains and transmission gearing which might give trouble, and differing from the earlier model flying boat built in San Diego, California, last winter (1911–12), which was equipped with "tractor" propellors—propellers in front—driven by chains.

The new flying boat is twenty-six feet long and three feet wide. The planes are five and a half feet deep and thirty feet wide. It runs on the water at a speed of fifty miles an hour, and is driven by an eighty horse-power Curtiss motor. At a greater speed than this it cannot be kept on the water, but rises in the air and flies at from fifty to sixty miles per hour.

The boat itself is provided with water-tight compartments so that if any one compartment should be damaged the flotation afforded by the other would be sufficient to keep the craft afloat. It is also provided with wheels for making a landing on the shore; these wheels fold up, thus not interfering in the slightest with its manœuvres over the water. The boat is so strongly built

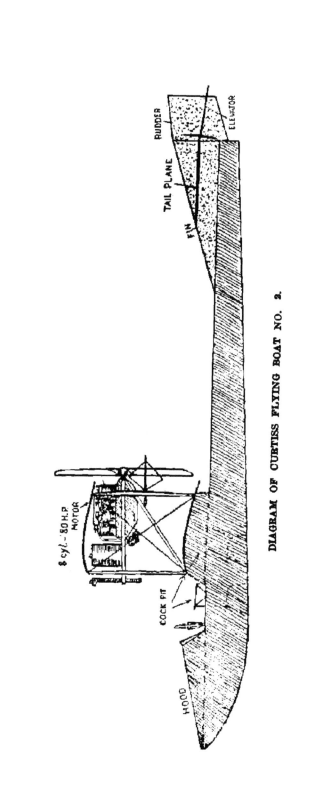

DIAGRAM OF CURTISS FLYING BOAT NO. 2.

that it can be readily beached even through a high surf and handled the same as a fisherman would handle his dory, or it may be housed afloat like a motor boat or anchored to a buoy like a yacht.

In rough water the spray-hood with which this type of boat is provided protects the navigators from getting wet and enables the craft to be used very much as you might use a high speed motor boat, with the added excitement of being able to rise above other crafts or fly over them if they get in the way. It looks very much like a flying fish in the air and although designed to skim close to the surface of the water at high speed it can rise to as high an altitude as the standard land machine.

Mr. Curtiss states: "My idea was to provide a machine especially adapted for the requirements of the sportsman, one that would be simple to operate and absolutely safe. During the tests which we have made with this flying boat it carried three people with ease and the boat rose without difficulty with the extra passenger, although it is only designed to accommodate two people."

With the hydroaeroplane a safe landing can always be made, and if, through inexperience or carelessness of the driver, a bad landing is made, no injury to the operator or passenger can occur other than what may result from a "ducking."

This boat shows how directly aeroplane-build-

THE EVOLUTION OF THE HYDRO

(A) (B) The flying boat of summer, 1912—on land and in the air. (C) A contrast—the hydroaeroplane of winter, 1911

HYDROAEROPLANE FLIGHTS
(A) Curtiss driving the "Triad" over Lake Erie. (B) Witmer riding the ground-swells at Atlantic City

ers are turning to air craft available for amateur sport—not for exhibition "stunts." Such boats will have ample protection for the passenger and be able to carry a large quantity of fuel together with wireless apparatus and provisions, so that long overwater journeys may be made in comparative comfort, and also well within the radius of communication by wireless. And most of all —they are safe!—A. P.

THE NAVY ON THE HYDRO
(AUGUSTUS POST)

Captain Washington Irving Chambers, head of the Aeronautical Bureau, United States Navy, in a speech delivered at the Aeronautical Society's banquet in New York, said:

"The hydroaeroplane is the coming machine so far as the navy is concerned; in fact, it has already come.[1] The navy machine built by Glenn

[1] The fame of the hydroaeroplane has reached the Orient and a demonstration was recently given at Tokyo, Japan, for the benefit of the Japanese Army and Navy officials by Mr. W. B. Atwater, of New York. Mr. and Mrs. Atwater are on a tour of the world, carrying with them two Curtiss hydroaeroplanes and giving demonstrations of a practical character before the military authorities of all the countries en route. On Saturday, May 11th, 1912, he made three flights at Tokyo, the first hydro flights ever seen in the Orient.

There was a great gathering of military men to witness the flights, among them Prince Kwacho, representing the Japanese Imperial Family; Admiral Saito, Minister of the Imperial Navy, and Vice-Admiral Uryu.

Curtiss has had several tryouts and has proved itself a success. I recently had a flight with Mr. Curtiss in this machine, the 'Triad,' at Hammondsport, N. Y.

"With two passengers seated side by side, the control can be shifted from one to the other easily while in the air. When we had gone a mile Curtiss yelled to me to take the control. The levers had been explained to me on the ground, but I had not familiarised myself with them for the purpose of handling the machine under way. I turned on a notch and the front plane tilted up, bringing the machine off the water to a level of four feet in the air. We kept this level for another mile or two, when Curtiss took the control again. He turned the plane lever another notch and we rose to a ten-foot level and encircled the lake several times without changing from this level more than a foot or two, lower or higher."

As a justification of Captain Chambers' remarks, the Aero Club of America, at their annual banquet held on January 27, 1912, awarded the "Collier Trophy" to Mr. Curtiss for his successful development and thorough demonstration of the hydroaeroplane, the terms of the deed of gift stating that "it shall be awarded annually for the

According to the statement of the *Japan Advertiser* the Japanese Navy has followed the example of Russia, and forwarded to America an order for four Curtiss hydroaeroplanes.—A. P.

greatest achievement in aviation in America, the value of which has been demonstrated by use during the preceding year.''

The trophy is a group in bronze by Ernest Wise Keyser of New York, representing the triumph of man over gravity and other forces of nature. The trophy was donated by Robert J. Collier, president of the Aero Club of America.—A. P.

PART IV

THE REAL FUTURE OF THE AEROPLANE

BY

GLENN H. CURTISS

WITH CHAPTERS BY

CAPTAIN PAUL W. BECK, U. S. A., LIEUTENANT
THEODORE G. ELLYSON, U. S. N., AND AUGUSTUS POST

CHAPTER I

AEROPLANE SPEED OF THE FUTURE

IF you look over the books on aviation that were published even a comparatively short time ago, you will see how much of them is given to prophecies and how little to records of performance. Because, of course, as soon as the aeroplane came into existence every one with eyesight and a little imagination could see that here was a new factor in the world's work that would change the course of things in almost every way, and naturally every one began to forecast the possibilities of aerial flight. And at first, when the machine was really so little known, even to the inventor, that aviators hesitated to push it to the extreme of its possibilities, writers had more to say about what the aeroplane would probably do than what it had actually done. But the aeroplane, which is bound to break all speed-records, has made history at the fastest rate yet. Day by day we move things over from the prophecy department to the history chapter, and as the days slip by on their rush to join the future, hardly one but leaves a record of accomplishment and achievement to justify the aeroplane prophets.

At first, as I have just said, aviators could not believe in the powers of the machine; we used to trim down our garments to the lightest point, to avoid extra weight, whereas now we bundle up in heavy furs, or wear two suits, one over the other, to meet the intense cold of the upper air; and a great surplus of weight can be carried by almost all machines. We used to wait for a calm almost absolute before going up—it used to be a regular thing to see aviators wetting their fingers and holding them up to see from which direction the faint breezes were coming—or dropping bits of paper to see if the air was in that complete stillness we used to think necessary for successful flight. When I was waiting for just the right moment in Albany to begin the Hudson Flight—which, because of the unusual and absolutely unknown atmospheric conditions over a river flowing between precipitous and irregular hills, had to be timed with unusual care—the Poughkeepsie paper in an editorial said the "Curtiss gives us a pain in the neck."

Even after I had made the flight the *Paterson Call* made the wait a reason for denying the use of aeroplanes in time of war, pointing out how amusing it would be to see in the newspaper reports of the wars of the future, "Battle postponed on account of the weather!" Whereas now we go up without hesitation into what is actually a gale of wind, and under weather conditions that would have made the first flyers think it absolute suicide.

This discussion of the future of the aeroplane will have more of a basis of solid fact for its prophecy than if it had been written a couple of years ago. Some ideas the world has as to the future of the machine we have had reluctantly to abandon—or at least indefinitely to postpone, but so many new fields of activity have opened that one may only sketch the principal lines along which it is reasonable to expect the aeroplane and the art and science of mechanical flight to develop.

The most practical present and future uses of the aeroplane in the order of relative importance which it seems to me that these uses will naturally take, are: for sport, war, and special purposes which the aeroplane itself will create.

SPEED—PRESENT AND FUTURE

In saying "for sport" I mean both for the aviator himself and for the spectators interested in watching his aerial evolutions and enthusiastic over results; over sporting competitions, speed races, and record flights of all kinds. Such flights provide as much fun for the fellow who looks on as the fellow who flies and gives an opportunity for those who take pleasure in acting in an official capacity to exercise authority to their hearts' content!

Speed will always be a most important factor in the development of the sporting side of aviation. Almost all races depend upon speed and activity; and the aeroplane, the material embodiment and symbol of speed, equals and in many cases surpasses the speed of the wind.

Speed will have no bounds in the future. As I have already said briefly in passing, aeroplanes will soon be going considerably over one hundred miles per hour. A motorcycle has gone at the rate of one hundred and thirty-seven miles per hour and an aeroplane should be able to go even faster. With the help of a strong wind blowing in the direction of flight, two hundred miles an hour ought to be possible of attainment. Machines for high speed, however, must have some means of contracting the wing area or flattening out the curve in the planes so that when we want to go fast, we can reduce the amount of surface of the machine to lessen friction and so that when we want to go more slowly and land, we can increase the size of the wing surface.

The Etrich machine built in Austria has been constructed so that the curvature of the planes can be changed by operating a lever near the pilot; this enables the machine to attain high speed in flight and to fly more slowly in starting and landing.

The record is one hundred and eight miles an hour now (September, 1912) and we will not be surprised to see it climb up in proportion as rapidly as the altitude record did in 1911.

There is no wonder that an aeroplane race should create such absorbing interest, almost amounting to a craze, in the mind of the public directly interested. Speed is the one thing about the aeroplane that appeals both to the practical and to the imaginative man; the man of business, to whom saving time means saving money, and the poet, or the man of leisure, to whom the words "make a bee-line"—that is, an air line—have always stood for speed and directness. Now in earth or rail friction-machines, the limit of speed has almost been reached, except in the case of monorail vehicles, and there seems to be little progress in this direction. With the aeroplane, on the contrary, speed is only in its infancy. None of the difficulties that check the development of speed in the automobile or locomotive attend the aeroplane. What means speed now—ninety or ninety-five miles an hour—merely marks a stage in the machine's development; a hundred and fifty an hour is even now within its possibilities, and a much greater speed is by no means beyond the vision of the present generation. What the boys of to-day are going to see when they grow up no one can foretell. It is largely a question of motive power—that and the reduction of resistance.

In the latter respect I have already materially cut down the resistance of the newest type of Curtiss machine, in order to increase the speed. I was able, as I have said, to win the International Cup at Rheims in 1909 with a speed of forty-seven and one-half miles an hour. At Los Angeles during the past winter my latest type was able to fly more than seventy miles an hour, and the same type of engine, an eight-cylinder, has also been made more powerful, thus the increased speed is due to the improvements in the lines of the machine, the reduction of surface, and the controls, and the increase of the power of the motor.

There is still room for reduction of surfaces, minor improvements in the general outlines and in the control; but the largest element in any increase of speed must rest with the development of the motor. Increased power is the tendency, with as much reduction in weight as possible. Personally, I can't see much room for reduction in weight. At present I am using a motor of my own manufacture that weighs but three pounds to the horse-power. This I consider extremely light as compared, for instance, with the engines used in submarines of the Navy, which weigh from sixty to seventy pounds to the horse-power. Still, there will be some reduction in weight per horse-power.

With the great speed that will undoubtedly mark the aeroplane flights even of the near future,

FUTURE SPEED 161

the physical endurance of the operator will count for a great deal in long flights. By the time we can fly much over a hundred miles an hour there will have to be some means of protection devised for the operator, for anyone who has travelled sixty or seventy miles an hour in an automobile knows how uncomfortable such a trip becomes if it keeps up over long distances. The driver of an aeroplane sitting out in front unprotected causes far more "head-resistance." It will be an easy matter to arrange some sort of protection for him.

How strong this "head-resistance" can be, I realised in a curious experience while racing with Ely at Los Angeles, going at probably sixty-five miles an hour. I looked upward to see just where Ely was flying, and as I raised my head the wind got under my eyelids and puffed them out like toy balloons. For a moment I was confused and could scarcely see, but as soon as I turned my gaze on the ground the wind pressure forced the lids back into their normal position.

SAFER THAN AUTOMOBILE RACES

I believe there are fewer dangers in racing aeroplanes than in racing automobiles. Races run over the ground have to contend against obstructions to the course, tire troubles, and "skidding" on a wet track, or in making sharp turns. None of these exist in the race in the air. The course is always clear, there is no "track," wet or dry, and as for the turns that look so desperate to the inexperienced observer on the ground, the operator, far from slipping out of his seat as he "banks" sharply, sits tight and feels as if he were going on an even keel. If you can imagine how the water in a pail would feel as you swing the pail around your head so fast that not a drop spills, you can realise the sense of stability that the aviator feels as he whirls around a circular course at a tremendous rate of speed, in fact, once an aeroplane is up in the air, it is often safer to travel fast than it is to travel slow.

ACCIDENTS

Of course it would be folly, in view of the list of accidents, fatal and otherwise, that the newspapers print and reprint every time a noted aviator falls, to assert that there is no danger in flying. I doubt if the American man, especially the American young man, would take to the aeroplane so enthusiastically if the sport were as safe as parlour croquet. There is, of course, always danger of something going wrong with an aeroplane in flight that may bring it down too quickly for safety, but unless the derangement is vital, an expert aviator can make a safe landing, even with a "dead" motor. And the dangers of flight are growing less and less every year as the machine is improved and as the aviator becomes more skilful and more experienced in air conditions. The report of the French Government for 1911 shows that there have been only one-tenth as many fatal accidents in proportion to the number of flights made, as in the first year of aviation, but each accident has made ten times as much stir.

INCREASE IN SKILL

Perhaps the greatest advancement in aviation during the past year has been due to the increased skill of the aviators. Men like Beachey, McCurdy, Willard, Brookins, Parmelee, Latham, Radley, and others who have made flights in this country, have shown remarkable strides in the art of flying. This advancement has been in experience—in knowing what to do in all sorts of weather—in taking advantage of air currents and in knowing how to make safe landings when trouble occurs. A year ago it would have looked like a desire to commit suicide to attempt some of the "stunts" these men now perform as a part of their daily exhibitions.

At the same time, I want to make it plain that, personally, I do not now, nor ever have encouraged so-called "fancy" flying. I regard some of the spectacular gyrations performed by any of half a dozen flyers I know as foolhardy and as taking unnecessary chances. I do not believe fancy or trick flying demonstrates anything except an unlimited amount of nerve and skill and, perhaps, the possibilities of aerial acrobatics.

CROSS-COUNTRY RACES

The year 1912 in America is the year of great cross-country flights. We have already seen the foreshadowing of this development in the great flights of Atwood from St. Louis to New York and Rodgers from coast to coast. Rodgers' trip was a great feat. Just think! Clear across the United States and so many smashes that only a man with indomitable will and pluck would have kept on to success. Rodgers became an expert at landing and made landings almost anywhere. Soon we shall see, instead of men flying alone as in the case of these trips, double flights with two pilots relieving each other so that the distance covered in flights may be increased, and the capabilities of machines for endurance can be fully shown.

The Gordon Bennett International Cup race at Chicago this year brought to this country two of the best racing machines in the world and has stimulated interest in aviation to a higher pitch than it has ever had in the United States. At the next Gordon Bennett, I hope to see an American surpass even Vedrines' speed of one hundred and five miles an hour for one hundred twenty-four and eight-tenths miles.

RACING TYPES OF TO-MORROW

There have been many meets since Rheims, some international, some of local importance; indeed almost every citizen of a civilised country has had a chance to attend some one of them without too great a journey, but what I have said of one meet is true to some degree of all: that racing and contests in general, especially between different makes of machines, is of the greatest use to the development of the aeroplane, just as competition among automobile manufacturers, in putting out racing machines, helped the development of that vehicle.

There are at the present time a number of types and makes of aeroplanes, each claiming some especial advantage over the others, and trying to demonstrate it. Some of these will drop out—some of them have dropped already—some will develop toward the aeroplane of the future, which we can only infer from the machines of to-day. The way to bring about this "survival of the fittest" is by speed contests and endurance races, where the American manufacturer pits his machine against the foreign-made article and the biplane contends against the monoplane.

The public believed, when these two types came into being, that there would be a sharp division of uses between them; that the biplane would excel in just certain directions, the monoplane in others, and the public has watched the various records of speed, of endurance, of distance, as they changed back and forth between the two types, and has found that deciding their relative merits and assigning their special uses was by no means the simple and summary process they thought it would be. The contests will have to evolve new rules and regulations; for instance, there will have to be some means of handicapping machines with very high-power engines and small plane surface—as in the case of monoplanes, which, with a minimum of plane surface and high power engines, have a speed advantage over the biplanes, that with equal engine power have much larger plane surface. Perhaps the method of handicapping now used in certain races of stock automobiles, that is cubic displacement of the engine, will be adopted.

PUBLIC INTEREST IN MEETS

The aviation meet at Los Angeles, California, in 1911, was a good indication of what great and deep interest the public have in contests in the air, and will have in the great races of the future.

Aeroplane flights called thirty thousand people through the gates the second day of the ten days' meet. This is the biggest crowd, I believe, that ever paid admission to an aviation meet, in this country, and probably the largest that has ever attended any outdoor attraction except the world's series baseball games and the few big football games. In addition, there was a considerable crowd on the outside who did not pay admission, but the actual paid admissions on Sunday were more than thirty thousand. This third annual meet did better than either of those held during the two previous years, and this, I am convinced, proves that aviation is a standard and lasting attraction.

CHAPTER II

FUTURE SURPRISES OF THE AEROPLANE—HUNTING, TRAVEL, MAIL, WIRELESS, LIFE-SAVING, AND OTHER SPECIAL USES

MANY will be the future uses of the aeroplane; special uses not necessarily dependent on speed.

Sportsmen are likely to find in the aeroplane, especially in the hydro, an admirable vehicle for hunting, aside from their interest in its racing capacity. Already there is pending in the California legislature a bill designed to regulate shooting from an aeroplane, intended as an addition to the California aeroplane traffic regulations, described later. While this bill is probably intended as more or less of a joke, it has been thoroughly demonstrated that it is possible to shoot wild ducks from an aeroplane. Hubert Latham proved this fact in his Antoinette monoplane at Los Angeles.

Latham flew from Dominguez Field to the Bolsa Chica Gun Club on the shore of the Pacific, ten miles away, and chased wild ducks for thirty minutes, finally bagging one. The sportsmen of California thought they saw in this feat of Latham's the near approach of a time when the aeroplane would be utilised for exterminating game, and seemed much exercised over the incident. The newspapers saw only the humour of the incident, however, and the sportsmen were quickly reassured.

Latham, not content with this achievement and thirsting for new thrills, said that he was going to fly up into the Rocky Mountains and shoot grizzly bears. His last undertaking was to take his aeroplane with him to the Congo where he went to hunt big game and to use the aeroplane in this novel and sensational sport. Strange to relate, after having braved all the dangers of the air, he met his fate by being gored to death by a wounded and infuriated wild buffalo, in July, 1912.

Some ranchers out west have clubbed together to purchase an aeroplane for hunting wolves which have been killing their cattle, and four aviators flew over San Fernando Valley in California recently, eagerly watching the underbrush for a sight of two fugitive bandits who for two days had eluded a large sheriff's posse after attempting to hold up a railway agent and mortally wounding a deputy at San Fernando. Each aviator was sworn in as a deputy and carried with him an observer provided with a powerful field glass. They reported that they could see objects very clearly below.

In scouring the hills one of the observers thought that he had surely spotted his man and the plane was dipped abruptly toward the ground. On returning he said, "It was a dog I saw and I'll bet that dog is running yet."

I have heard on the best of authority that an aviator in this country chased a buzzard until it fell exhausted and that in Europe this same game was played by a German aviator upon a large stork.

AERIAL BIRD-NETTING

On my practice flights in a hydroaeroplane over San Diego Bay, I noticed on several occasions that pelicans and sea gulls and even wild ducks got in my path, and I was sometimes obliged to change my course in order to avoid the slow-flying fowl. It occurred to me that with a net affixed to the forward part of the planes it would have been an easy matter to run down and bag a pelican, and possibly a sea gull. The ducks are too quick to be caught by an aeroplane, as yet. Chasing ducks in an aeroplane and catching them in a net would be about as thrilling a sport as one can imagine. Perhaps when the killing of wild fowl with guns shall have palled on sportsmen, we shall see the method of "netting" them with an aeroplane come into use. Something after the manner of scientists who hunt the *lepidoptera*.

AEROPLANE SURPRISES   171

Mrs. Lillian Janeway Platt Atwater, of New York, while taking instructions in the operation of the hydroaeroplane at North Island, early in 1912, tried my new method of catching seabirds. She asked Lieut. J. W. McClaskey, instructor at the Curtiss school, to take out the hydroaeroplane, with her as a passenger, and attempt to catch a pelican or gull with a net. The instructor promptly agreed and for almost half an hour the big hydroaeroplane with Lieut. McClaskey and Mrs. Atwater chased pelicans and sea gulls up and down the bay. They discontinued the hunt only when a large pelican barely escaped becoming entangled in the propeller, which would have smashed it and possibly caused an accident. On another occasion Mrs. Atwater did actually succeed in catching a gull while flying with her husband.

Shooting rabbits from an aeroplane would be comparatively easy. I came to this conclusion while flying over North Island, which is covered with weeds and sagebrush for the most part, with hundreds of jack-rabbits and cottontails living there. At first these rabbits were terribly frightened by the aeroplane and ran in all directions to escape. They soon became used to the sight, however, and would watch the aeroplane with a great deal of curiosity. One of the big jack-rabbits, either from fright or curiosity, waited too long to get out of the way of Harry Harkness in his Antoinette, when he made a rather abrupt descent, and it was cut in two by the propeller.

MAIL-CARRYING

One of the most important special uses to which the aeroplane is particularly adapted is for carrying the mail. Royal mail was first actually handled at Allahabad in India last summer, during which over 6,000 letters were transferred. This service was planned to prove the great value of an aeroplane post during war time to a besieged town. A mail route via aeroplane was established on trial between London and Windsor in England, which carried several tons of mail matter. And in this country last fall Postmaster-General Frank H. Hitchcock and Captain Paul Beck, U. S. A., inaugurated the first aerial postal service regularly established in the United States, over a route between the Aero Club of America's flying grounds at Nassau Boulevard on Long Island, and Mineola, L. I. A picturesque account of this little episode is given by Frank O'Malley, who wrote:

"The flying events of the day at the Nassau Boulevard aviation meet came to an end in a hubbub of joyousness among 1,500 spectators on the grounds.

"Lieutenant Milling had busted the American record and was still flying for the world's record when a tall, youngish man decked out in a blue serge suit, and a gray cap, climbed into the Curtiss machine driven by Captain Paul Beck of the army.

" 'The Hon. Frank H. Hitchcock, Postmaster-Gen'rul of the whole United States,' the megaphone man began to holler, 'will now fly to Mineola with Captain Beck to deliver the mail. Postmaster-Gen'rul Hitchcock of the United States will carry the mail-bag on his knees and drop the bag at Mineola into a circle in which will be the Postmaster-Gen'rul of—I mean the Postmaster of Mineola. Ladies and gentlemen, Postmaster-Gen'rul Hitchcock.' (Much applause.)

"Mr. Hitchcock wasn't around to hear all this and so didn't lift his gray cap in acknowledgment. He was far out on the field with Attorney-General Wickersham and Captain Beck. Post Office Inspector Doyle handed the Postmaster-General a mail bag containing one thousand, four hundred and forty postcards and one hundred and sixty-two letters, and Captain Beck and the Postmaster-General hiked off in a northerly direction for the high spots.

"The Curtiss circled three-quarters of the field and then climbed rapidly until it was three hundred or four hundred feet above the south end of the track. Ovington, who had also got under way with a second bag of mail in his monoplane, shot up into the same acre of sky occupied by Captain Beck and Mr. Hitchcock and shot eastward as a track finder for Captain Beck's machine.

"The field could see the two machines almost all the time during the cross-country flight. The way the biplane with a passenger pegged along just behind the monoplane with only a pilot aboard was a caution. Over a big white circle painted on the Mineola real estate, Ovington from his monoplane and the Postmaster-General from Captain Beck's machine, plumped down to Mineola the two pouches and hit within the circle in each case.

"The biplane teetered slightly as the mail bag was released and then the two machines made a circle and spun back to where the crowd stood on tiptoe peering over fences at Nassau Boulevard.

"'I was up once before,' the Postmaster-General said after he had shaken hands all around upon his return to earth. 'That was at Baltimore with Count de Lesseps in his Blériot. The biplane to-day I found was much steadier.

"'Fly again? I hope so, because I like the experience very much. My trip to-day was especially enjoyable because at Baltimore I could see very little of the ground below, owing to the closed-in construction of a monoplane. To-day from the biplane all this end of Long Island was stretched out to be looked at.

"'Yes, air-routes are all right for practical

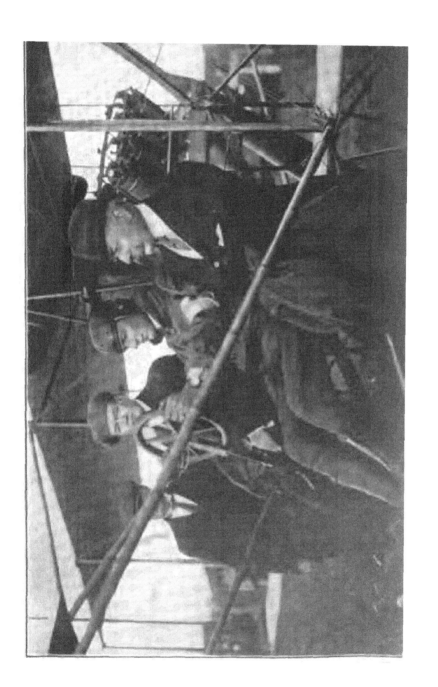

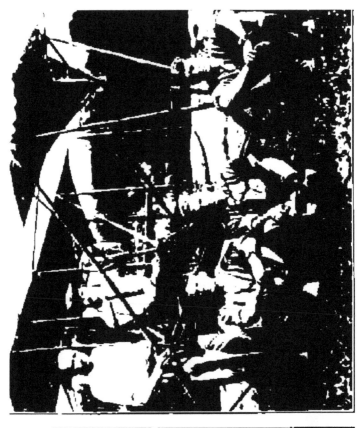

mail-carrying,' Mr. Hitchcock continued, in answer to a question. 'I mean,' he smiled, 'the air is all right, but the vehicles must continue toward perfection. But even with the aeroplane as it is now it would be very useful to us, particularly in some parts of the country.

PRACTICAL VALUE TO-DAY FOR MAIL-CARRYING

" 'Take along the Colorado River in the cañon district of Yuma, for instance, or in parts of Alaska. Along the Colorado there are places where detours of fifty miles out of the way are made in mail routes to get to a bridge. An aeroplane could hop right across the river.

" 'The expensiveness of maintaining an aeroplane service is an obstacle, but that will diminish. I would like to see the Post Office Department do something definite in this direction for the good effect it would have in stimulating the development of the machine. Fliers at present have many lean months between the meets.' "

Ever since Postmaster-General Hitchcock made this trip he has been an enthusiastic advocate of the aeroplane as a means of transporting mail over difficult routes. During the next few months he granted permission to a number of aviators, including Ovington, Milling, Arnold, Robinson, Lincoln Beachey, Charles F. Walsh, Beckwith Havens, Charles C. Witmer, and Eugene Godet, all of whom fly Curtiss machines, to act as special mail carriers, and these men have carried mail bags in similar exhibiting tests from aviation fields to points near the Post Office. Among the cities where such tests have been officially made are Rochester, N. Y.; Dubuque, Iowa; Fort Smith, Ark.; Temple and Houston, Texas; Atlanta, Savannah, Columbia, and Rome, Ga.; and Spartanburg and Salisbury, N. C.

The record for long-distance mail carrying is held by Hugh Robinson, who took a bag of mail at Minneapolis, Minn., and carried it on his long flight down the Mississippi River in a hydroaeroplane as far as Rock Island, Ill. The distance covered by Robinson was 375 miles on this trip, and letters and first class mail matter were put off and taken on at Winona, Minn.; Prairie du Chien, Wis.; Dubuque and Clinton, Iowa; and Rock Island, Ill.

Of course the aeroplane is, at present, best suited for carrying mail in localities where the weather is equable; in such places it offers a speedy, direct, and dependable service. These numerous experiments in mail-carrying by aeroplane have brought about the urging of an appropriation by Congress for this purpose. The second Assistant Postmaster-General, who is in charge of mail transportation, in a report that has just been made public at the time I am writing this, asks for $50,000 for the transportation of mails by aeroplane. Part of this fund may be devoted to mail routes in the Alaskan interior.

One government has actively entered on practical mail-carrying by aeroplane. Belgium has voted a fund to establish routes across seven hundred miles of impenetrable Congo jungle.

WIRELESS

The aeroplane is ideal for use with wireless telegraphy and the combination of the aeroplane's ability to obtain information and the ability to transmit it by wireless will be one of its most important future developments in practical usefulness.

Wireless experiments do not involve any great problem, as messages have been successfully transmitted from an aeroplane to land stations many times. The receiving of a wireless message by an operator in an aeroplane from a land station or from a warship involves considerable difficulty because of the noise and vibration of the motor, but it is expected, however, that this will be soon entirely overcome and that it will be possible to transmit or receive telegrams in an aeroplane to or from distant points with the same ease and accuracy that it is now seen on the ground or on the water.

The telegraph seems to be the companion of the locomotive, the telephone of the automobile, and now wireless has its side-partner in the aeroplane!

Important experiments are being carried on by the signal corps of every army with various methods of communication with an aeroplane in flight and by the aviator with those on the ground. They have tried an instrument for making smoke signals, with large and small puffs, reviving a method used by the American Indians in the pioneer days and quite familiar to all boys who have played Indian in the country.

FORESTRY SURVEY

The supervisor of the Selway forest, consisting of 1,600,000 acres, which was formerly part of the Nez Perces reserve in Idaho, predicts that aeroplanes and wireless telegraphy will be important factors in forest fire prevention before a far distant date. He believes that a man in an aeroplane could do more accurate and extensive survey work in the forests of the Pacific slope country in a few hours when forest fires are raging than is usually accomplished by twenty rangers in a week. With wireless stations installed on peaks in the chief danger zones, he believed it would be a comparatively easy task to assemble men and apparatus to check and extinguish the flames and prevent fires from spreading.

MOVING PICTURES

Aeroplanes have already been used for purposes of photography and moving picture machines have also been attached to them and some remarkable pictures taken. One of the large moving picture magnates said, "Now, Mr. Curtiss, if you can take a series of moving pictures showing a trip across the United States, I do not care if it takes you a year to get it and even though it is taken piecemeal, or one section at a time over the main cities on the way, I will pay you well for it. We will take the film, trim it down, and run it through at lightning speed taking our audience from New York to San Francisco 'as the bird flies' in twenty minutes."

The value of moving pictures taken from above —and from a swift low-flying machine—is apparent at a glance. The contour of the country is shown as in no other way, and now that warfare is going to have a quite different point of view, even a different range of action, it is important that schools, and especially military schools, should be made familiar with this aspect of the land. The flat map is superseded by such a panoramic view. In time of actual war, moving pictures taken in this way will have a unique value.

In photographing reviews of troops, public celebrations, lines of battleships, or any scenes that require a panoramic representation, the aeroplane has been used with success. It can also be of great service in photographing animals and rare birds which may inhabit regions otherwise inaccessible. With the advance of nature study and the steady development of "camera hunting," the aeroplane will be used more and more for such purposes as well as for photographing mountain tops and other insurmountable or dangerous places to reach.

Robert G. Fowler has had some surprisingly good motion pictures taken from his machine during his cross-continent flight, by an operator sitting beside him, his camera placed on a temporary stand.

Mr. Frank W. Coffyn took a most interesting series of moving pictures of New York City from the water front, portraying the Battery, the Brooklyn Bridge, and the famous Statue of Liberty in the harbour. Mr. Coffyn used a hydro-aeroplane for this purpose, which made his flights comparatively safe. In fact, such a feat would have been well nigh impossible for a machine that could not land on the water, for there are no places where an aeroplane can land in the business section of New York unless the aviator should land on one of the large buildings, and then he would have great difficulty in getting away again.[1]

Great care has to be exercised to keep the machine on an even keel, so that the operator can manage the roll of film.

[1] The first *start* from a roof-top was made on June 12, 1912, when Silas Christoferson in a Curtiss biplane rose from a platform built on the roof of the Hotel Multuomah, Portland, Ore., and flew safely away.—AUGUSTUS POST.

LIFE-SAVING

Another branch of the government service that will no doubt be greatly aided by aeroplanes are the Life Saving Stations along the coast, whose regular equipment might well include an aeroplane to fly to wrecks and carry a line from shore to ship when the high seas make it impossible to launch a lifeboat. It might be impracticable to go out during the period of severe storm, but there is always a calm in the air after a storm, as well as the proverbial calm before one, while the high seas in which a lifeboat cannot live are still running. The aeroplane or the hydroaeroplane, dashing through the air, even through high wind, would bring the line that means life to helpless men clinging to a wreck.

I am awaiting with earnest expectation the first time that an aeroplane actually saves a life; when that takes place, it will have conquered the heart of the people as well as fascinated its intellect, aroused its awe, or compelled its admiration. The first period of enthusiastic acceptance of the new machine has been succeeded in the mind of the general non-flying public by an admiration not at all like affection.

Realising how many lives have been given to its development, feeling that the aviator takes, as they call it, "his life in his hands," the crowd at a flying-meet feels with all its great and growing interest, an attraction in which figures not a little fear and distrust. The first time that an aeroplane saves a life—as it can and will do many times—it will have begun to conquer this public distrust. That is why the exploit of the hydroaeroplane already described, in coming first to the aid of the aviator in the water, had a value far greater than its apparent importance.[1]

[1] A very important service was rendered only a short time ago by the hydroaeroplane which might easily have served to save a human life if the accident had been more serious than it actually was. Mr. Hugh Robinson the instructor of the Curtiss hydroaeroplane school was having Sunday dinner at the hotel in Hammondsport, where Dr. P. L. Alden, one of the well-known physicians of that place, was also eating dinner, when the doctor received a telephone message that Mr. Edwin Petrie's little son had fallen from the steps of the Urbana Wine Company at Urbana, five miles down the lake, and had a compound fracture of his thigh with a serious hemorrhage. It was a very serious injury and the little fellow was in intense pain, and Mr. Petrie asked the doctor to come as quickly as he possibly could.

Dr. Alden realised the urgency of the situation and knew that delay might mean serious results from hemorrhage, so he went immediately over to Mr. Robinson and asked if he would take him across the lake in the hydroaeroplane right away. Mr. Robinson said, "I will be ready in five minutes; just as soon as you can get over to the field."

Dr. Alden got his bandages and instruments and hurried down to the shed where Mr. Robinson had already gotten out the hydro; he jumped in and they were off without a moment's delay. They covered the five miles in five minutes, at times running on the surface of the lake because the wind was blowing so strong; as they ran up on the beach the doctor jumped out and hastened to his patient.

The boy was so much interested in the fact that he was the first

EXPLORING AND ESCAPE FROM DANGER

The aeroplane will find one of its important uses not only in taking pictures of inaccessible spots, but also in crossing otherwise impassable places, especially in times of pressing need when fire, earthquake, volcanic eruptions that leave beds of molten lava, explosions, pestilences, floods, or other devastations occur, and quick assistance is necessary.

In engineering and mining matters, the aeroplane may be of assistance in exploring the best places to locate the route for railroads through mountain passes and into such places as "Death Valley" where the salt deposits are located.

TRAVEL

An important field in the sporting world of aviation of course will be carrying passengers and initiating novices into the mysteries of the air lanes and into the pleasures of aerial touring.

In this delightful method of travel the panorama below is equal to any of the magnificent

patient to be treated by a hydroaeroplane doctor, and so fascinated at hearing Dr. Alden tell about the trip, that he forgot for the moment the seriousness of his condition and allowed the doctor to reduce the fracture without an anesthetic. When all that could be done just then had been done, Dr. Alden and Mr. Robinson returned in the hydroaeroplane as rapidly as they had come on their errand of humanity, and at last accounts young Mr. Petrie was getting well as fast as he could so he could have a ride in the hydroaeroplane himself!—AUGUSTUS POST.

landscapes which may be seen from high mountains and besides, the view is attended by most delicious thrills and sensations, and when a good pilot is in control of the machine the passenger is sure of a pastime absolutely unequalled for mere joy, aside from further use or benefit it may have.

While travelling over torrid places like deserts and arid wastes, as well as burning prairies, the aviator can fly high where the air is cool and clear and escape the great humidity and the deadly alkali dust.

As for mountain climbing, it will have lost its peculiar fascination when the aeroplane will be to mountains what the elevator is to high buildings. The landscape has a greater, far greater beauty; for an aviator can see a great distance over a level plane. At the height of one mile you can, theoretically, see ninety-six miles in every direction and as you ascend the distance to the horizon becomes greater. In hilly country, one hill hides another when you look from the ground, but when you are high up in the sky, like the eagle, the mountains all seem to lose their height and appear flat and naturally your view is unobstructed.

At great altitudes the sky becomes very deep blue and if you kept going up you would reach a point finally where the sky became black and the sun appeared like a ball of fire all by itself as a candle flame does in the dark.

FOR HEALTH

In these regions there is no dust in the air to diffuse the light and the air is dry and consequently excellent for persons with lung trouble. There is even a possibility that physicians will advise patients suffering from tuberculosis to ascend to these high altitudes, and it is a fact that Hubert Latham was threatened with this disease, yet enjoyed good health after taking up aviation, only to be killed by a wild buffalo, as related. Perhaps this is one of those cases I was looking for where the aeroplane has saved a life.

METEOROLOGY

An aeroplane will bring quick reports of changes in the weather. Rapid investigations of conditions which exist in the strata of air at varying altitudes above the surface of the earth, made by the use of flying machines, may lend us material aid in understanding those conditions which are closer to earth.

The study of the weather and meteorological conditions becomes of greater and greater importance as the progress in the science of aviation advances. The currents of air that are regular in their direction of movement, like the trade winds, must be mapped and charted, for with the aid of a strong wind an aviator can make marvellous speed, as the speed of the wind is added to the speed of his machine and with an aeroplane capable of making one hundred miles an hour a favourable wind of fifty miles an hour would increase the total speed by one half. For the wind is now no longer an obstacle to flight, and as I have already noted at the beginning of this chapter, this is one of the most noticeable advancements in aviation, one that can readily be seen, understood, and appreciated even by the uninitiated.

THE TENDENCY TOWARD HYDROS

There is always more or less danger in flying over land, and the rougher its surface the more difficult and dangerous the matter of landing. The safest place and the most uniform surface is to be found over the water, and there is much less danger to the aviator flying there than over the land. The strength of the wind can be easily judged by the height of the waves, and squalls and puffs can be seen coming so that if they seem to be very bad you can come down on the surface of the water or skim along very near it with the greatest safety, if you are in a hydroaeroplane. Rivers will no doubt become the favourite highways of travel for the airman, as they were once the only great avenues for the march of civilisation when the canoe or the rude boat was the only vehicle of transportation. This brings us naturally to another consideration of the air-land-water machine.

CHAPTER III

THE FUTURE OF THE HYDRO

THE most interesting type of flying machine for sport and pleasure is the hydroaeroplane, and this is undoubtedly the machine with the greatest possibilities for the future. Indeed, it opens up an entirely new region of activity, as boundless as the ocean itself, and as various as the different bodies of water. Built along the lines of a motor boat with the addition of aeroplane surfaces or horizontal sails, this craft will be used for much the same purposes as motor boats are now, but in ways immeasurably more varied and more effective.

The boat portion will be made large and comfortable for pleasure trips and will be a veritable sportsman's machine which can be run up to a dock where it can make an easy landing and be tied up when not anchored out from shore. There will be a comfortable cabin, with cushioned seats for the navigators, and celluloid windows will be placed in the planes, so that the view below will not be obstructed. It will be handled in heavy seas without difficulty.

With such an air and water craft you can go off

hunting or fishing; you can shoot ducks and you will not have to wait until Mr. Duck comes by but you will be able to reverse the present custom and chase him in his native element and overtake him, too, as you would a fox on horseback. By rising to a good height you can see schools of fish or good places on the bottom to cast your lines for fishing.

Inland lakes will be just the place for the water machine and even among the mountains the surface of lakes will offer ideal places for landing and starting, even where the shores are quite out of the question for safe flying ground.

The construction of the hydroaeroplane, while keeping on the same general lines of development, will adapt itself to the exigencies arising from its extended uses. The propeller or propellers will be protected from the flying spray which might break them—for small drops of water are like bullets out of a gun when hit by the rapidly revolving blades of the propeller which travel so fast that water might just as well be solid matter as far as getting out of the way is concerned. Spray will chip pieces right out of a wooden propeller. Propeller blades are now covered with tin on the tips for use on the water, and even metal blades may be better in some respects for this purpose. The control and rudders are placed on the rear of the long light boat, which extends further to the rear to accommodate them.

The radius of action in the hydroaeroplane is now from four hundred to five hundred miles, for the machine can carry a barrel of gasoline, or fifty-two gallons, and as the engine uses about seven gallons an hour this would mean about seven hours running at from fifty to sixty miles an hour in still air; if the wind were blowing twenty-five miles an hour in the direction in which the machine was flying it would add two hundred and fifty miles to the distance covered in ten hours.

These machines can be equipped with more surface and they can be specially built to carry as much as two barrels of fuel, which would enable them to fly nearly twelve hundred miles if the wind were steady. They can also fly in very high winds up to almost one hundred miles an hour, which is the speed at which some of the higher air currents flow, as proved by the flight of balloons. This would of course tremendously increase the distance covered. All this is possible to-day and it seems that the aeroplane has already done every thing possible to be done on land. Blériot crossed the English channel, Chavez crossed the Alps, and Rodgers crossed the American continent, passing over the Rocky Mountains, and making over four thousand miles in the air.

The only thing now left is to cross the ocean. An attempt has been made to cross the Atlantic in a dirigible balloon. You all remember how

Walter Wellman flew out over the ocean from Atlantic City in one of the largest dirigible balloons ever constructed here, the "America," remained three days in the air, and covered over twelve hundred miles, even though his motors were running only a part of the time.

He was fortunate enough to be rescued and brought back to land by the steamer *Trent*. And nothing daunted, his chief engineer Melville Vaniman constructed another large dirigible the "Akron," on which he met such an untimely end.

Another entrant in the world race to cross the ocean is Dr. Gans who, with the backing of the German government, plans to start in his dirigible balloon the "Suchard" from the Island of Teneriffe, one of the Azores, to attempt the crossing of the Southern Atlantic. He will endeavour to be the "Columbus of the air" and be wafted above the waves by the selfsame winds which always blow in that part of the ocean to the West Indies, just as the first man to accomplish this passage was driven over the surface of the sea with his small ships. Such great enterprises bid fair to embolden aviators in their aeroplanes to try to win the laurels due the first to be successful.

Many aeronauts and aviators seem to be focussing their minds at the present moment on this great problem. It seems always a condition necessary to precede the accomplishment of any great

THE FUTURE OF THE HYDRO 191

thing that popular thought should be centred upon it; then some one rises to the occasion and the thing is done. There is no doubt that such a flight is possible to-day, just as the flight across the United States was possible in even the early stages of aviation. For the machine and motor which actually accomplished this trip were almost the same as the very first models; but it took the man to do it.

It will no doubt necessitate a double machine, and will need two pilots, one to relieve the other, and possibly several engines to ensure against stopping of the motor. Mr. Grahame-White has predicted that within twenty years we will be flying across the Atlantic in fifteen hours upon regular schedule between London and New York. Mr. Grahame-White once even went so far as to say that the ocean in a few years would only be used "to bathe in"—but I think he might have added "and to fish in," and left us that consolation!

Perhaps, backed by government aid, and with the co-operation of their naval vessels, a chain of ships could be stretched across the ocean, which would make it possible even now to fly with safety over the distance between Nova Scotia and Ireland, about two thousand miles. Already, Mr. Atwood who flew from St. Louis to New York, and Mr. James V. Martin, have seriously planned such a trip. Mr. Martin has submitted his plans to the Royal Aero Club of England. He proposes to keep in the track of steamers and to endeavour to secure the most favourable wind conditions possible. His machine is designed to have large floats and five powerful engines.

Storms pass across the ocean with great rapidity and a fifty-mile-an-hour wind would so increase the speed of an aeroplane as materially to help it on its journey.

The accomplishment of this great flight over the ocean will no doubt mean great things for the progress of the world but it also will require further development along the lines of a flying boat, where a substantial vessel will be provided, able to stand rough sea and yet able to rise and skim the surface of the water.

Following up the success of my new hydroaeroplane, I have taken great interest in the idea of a flight across the Atlantic Ocean by aeroplane. I consider the flight possible, and I am willing to undertake the construction of a machine for the purpose, provided any of the aviators now considering flight wish me to do so. I am not prepared to give the details of such a machine as would be required to make the flight, but I simply express the opinion that the feat is possible and that under certain conditions I will undertake to furnish the equipment.

CHAPTER IV

FUTURE PROBLEMS OF AVIATION

IN a consideration of the final structure of the Coming Aeroplane, we pass into the realm of pure prophecy, for the aerial liners and dreadnaughts of the future are still snug in the brains of men like Rudyard Kipling or H. G. Wells. My part in the consideration of what is coming is here confined to the consideration of the immediate, or at least the not far distant, future.

Biplanes will always be the standard machines in my opinion, because you can get more supporting surface for the same weight.

Surfaces may be set one far out in front of the other, as Farman has done, but with three surfaces the third requires a full set of struts and wires and just as much weight as for two ordinary surfaces, and adds only one half more surface, and the head resistance is also increased once again. Surfaces no doubt will be made larger and machines much bigger in every way will be built.

Telescoping wings may be a feature of the future machines, so that a graduated area of wing surface can be readily obtained and changed for slow or high speed.

The limousine, or enclosed-cabin body, will be a familiar sight in the future machines built for passenger-carrying. These cabins will be provided with comfortable seats.

AUTOMATIC STABILITY

In regard to the question of automatic stability, or some device to balance the machine automatically, there seems to be no doubt that this problem will be solved; in fact it is already solved both for balancing laterally and keeping the machine from tipping sideways and also to govern its fore and aft pitching.

These devices may be of value in learning to fly. But in the practical use of the aeroplane you may see conditions arising which you wish to counteract before they occur and for which you wish to prepare. Automatic stabilisers will no doubt prove very good auxiliary devices, and some aeroplanes will have automatic stabilisers on them before this is printed, but the aviator will no doubt have to regulate the regulators in the future as he operates the levers personally in the present.

AVIATION LAWS TO COME

The making of good laws is not to be overlooked when considering the future development of the aeroplane, for aviators must be protected from themselves, and the public must be protected from

FUTURE PROBLEMS 195

the rashness or inexperience of airmen. Almost all nations have already begun to exercise control over their new territory, the air, and are realising that it may become one of their most valued possessions and of an importance equal to their domain over water. For a nation without any seacoast may no longer be cut off from direct intercourse with the world through the aerial craft which can enter and leave at will, as vessels now do on the sea, with no chance of a neighbouring nation restricting this very freedom.

Laws are rapidly being passed by states regulating and licensing aviators and requiring lights to be carried, but it seems that the federal government should be the power that should control the air just as it does the sea and navigable rivers. For fliers flit about so that the whole country seems but a mere playground for men of the air.

Already the California legislature has made several laws to protect the aeroplane and the aviator, as well as to safeguard the larger public that stays on the ground. Some of these laws may seem a little premature, but everything about aeroplanes goes so fast, that there is no wonder the laws instead of lagging behind conditions as they usually do, should speed up a little ahead of them, for the progress of flight is such that by the time the law gets on the statute books the conditions may be calling for it. For instance, bills have been introduced at Sacramento to regulate the licensing of aeroplanes, which are to be classed as "motor vehicles," and to carry numbers and lights, the same as automobiles. The idea of providing for lights seems a little far-fetched at this time, as it will be a long time before there will be much flying at night. Besides, such lights as the proposed law provides would be unnecessary, for the reason that the aeroplane would not be confined to an arbitrary path, but could choose its own course. Therefore, a single light in front and another behind would be all that would be required, instead of one pair in front, one behind and one on each plane, as the bill suggests.

FUTURE COST OF THE AEROPLANE

The cost of the machine is high at the present time because there are but few made. No doubt when the great numbers of people who are now deeply interested in the subject get to the point of practical flight and desire to take flights, they will want to own machines, and learn to operate them. Then aeroplanes will be made in quantities and the price will be reduced in accord with the number that are built and some day we will be able to buy a good aeroplane for about the price we have to pay now for a small automobile.

Cortlandt Field Bishop is credited with having said when some one asked him if the manufacture of a cheap aeroplane, to cost $150, including the

FUTURE PROBLEMS

motor, would not be a great business undertaking, "Well, a great undertaking business should certainly come of it."

LANDING PLACES

The most serious problem of flying to-day is to find a good course to fly over and suitable landing places. The day will soon come when every city and town will have public landing and starting grounds. As a matter of fact the park commissioners of New York City have already been discussing the setting apart of landing places or isles of safety in the public parks of the city, although some authorities declare that it would not be well to encourage fliers to risk themselves and the people below by flying over the houses. There should be routes of travel established between cities over which an aviator will have a right to fly, just as there are highways on the surface of the earth.

GOVERNMENT ENCOURAGEMENT

Perhaps the greatest factor which is needed to further the development of the aeroplane to-day is the thorough appreciation by the National Government of the benefits which the aeroplane may bring to its various departments besides the military and postal service.

When railroads first became practical the government gave millions of dollars besides large grants of land to enable them to extend and develop to a successful state. Steamship building was helped in the same way both by government aid and by the building of warships and transports.

The French Government continues to lead the world in its encouragement of aviation. During the month of December, 1911, according to most reliable statistics, the War Department ordered no less than four hundred new aeroplanes, divided between a dozen or more types, and asked the government to appropriate the sum of $4,400,000 for aeronautics. Italy, next to France, is the most active European government in aviation, the Italian War Department having ordered fifty French machines of various types, as well as twelve aeroplanes of a new type produced in Austria. The Turkish government has decided to establish schools for the "fourth arm" immediately, while Russia will also increase its aviation programme. The latest government to take up aviation is that of Australia, where an aviation school is about to open for the instruction of army officers. Germany is not as active in aviation as the other principal European governments, although it is difficult to say exactly what is being done by the Germans, as they purchase machines made in their own country only.

A most interesting programme was arranged by the British military authorities for the trial of machines in competition in the summer of 1912, at

Salisbury Plain, in order to determine the best types of military aeroplane. The winning types in this contest will receive large orders from the British government to supply the Army and Navy with aerial equipment.

FIRST AVIATION REGIMENT
(Newspaper Despatch)

PARIS, Jan. 25, 1912.—The first aviation regiment, 327 strong, was organised here to-day.

A flag will be presented to the battalion later on.

Having already organised an aviation regiment, French army officers are now agitating the question upon the basis of having no less than a thousand aeroplanes ready at a moment's notice under the command of superior officers and under perfect control of army pilots trained to handle them. This training of officers is the most important part, for it takes time to make good fliers. Machines may be turned out very rapidly, but fliers become skilled to the point where they may be of use in army work only by long practice and practical experience. Our government has given an appropriation, small in comparison with what France, Germany and England appropriated, and we have a few aeroplanes in the signal corps of the Army now and three machines in the Navy, but these are only the first steps in this important branch of our military and naval development. We all hope for at least adequate equipment, an equipment that will equal, if not surpass, that of the European powers.

After the development of the aeroplane for sport and commercial purposes, its greatest field of growth is for purposes of war and here we find that the aeroplane can be at once the most deadly weapon of offensive warfare as yet developed by man, and an even more serviceable agent for defensive measures, or for all those most important duties related to scouting and obtaining and carrying information.

WHAT THE AEROPLANE CAN DO IN WAR

I feel confident that an aeroplane can be even now built which will be able to lift a ton of dynamite or other high explosive, and that it can be so constructed that it will be an aerial torpedo or winged projectile, the engine charged with compressed air and set to run any required distance, from one mile to ten miles. Such a machine can be steered by wireless controlling apparatus just as submarine boats and small airships are directed.

A hydroaeroplane can be made to fly at just a certain height over the water by attaching it to a drag or a float which would prevent its exceeding the desired limit of altitude. The machine so equipped might be started in a circle and flown around in a circular course gradually widening

FUTURE PROBLEMS

and widening, like a bird dog hunting a scent, until the object aimed at is hit.

One of the most important uses of an aeroplane adapted to the uses of the Navy will be its valuable assistance in enabling the manner of formation of the enemy's ships in line of battle to be made known to the commanding officer and the angle of approach to be estimated, in order that our own ships may be so formed in line of battle as to meet the brunt of the attack effectually.

An aeroplane launched from the deck of a battleship and ascending to the height of a mile will give the observers on board a range of vision of ninety-six miles in every direction and powerful glasses will reveal many details that can be seen more clearly from above than when observed from the same level. Submarines can be located with great ease when far below the surface of the water. Even the bottom appears clearly in some of the tropic seas, and fogs, which obscure all things to the enveloped mariner bound to the surface of the sea, usually hang comparatively low down and even a moderate altitude will enable an aerial observer or pilot to see clearly above the banks of mist which shut down like a pall upon the water.

The military aeroplane will be able to muffle its motor and for night operations will be equipped with search-lights and able to approach an enemy unseen and unheard from a high altitude, a direction in which there are no pickets.

In the school machines of one of the Chicago schools the motors have already been muffled to permit the teacher more readily giving his instructions to his pupils. U. S. Army officers have also experimented with mufflers on their motors.

Aeroplanes have been recently used by the Italian Army near Tripoli and bombs were dropped which not only frightened the enemy but stampeded their horses and caused panic among the soldiers. They were also of great service in directing the fire of the guns from the ships which were quite out of sight of their targets, a captive balloon and an aeroplane signalling the effect of the shots and the angles at which to train the guns. The aviators took steel bomb-shells with them and filled them while flying, holding the caps in their teeth, and steering with their knees while performing this operation. They did not dare to carry the bombs loaded for fear of being blown to pieces themselves in case of an accident when landing.

In the fall of 1911, extensive tests were made by the French military authorities which showed how reliable aeroplanes can be. The aviators flew at the command of officers and under the strictest orders; the machines were required to land in ploughed fields and to start away again with their full complement of passengers and extra weight of fuel. All the machines were required to carry a weight of about five hundred pounds and to rise to a certain height in a specified time with their complete load. The machines were also dismounted and assembled in the field and packed and transported from one place to another, to test the ease with which this could be done.

These military tests were won by Charles Weymann, who was also the winner of the Gordon Bennett International Aviation Cup for America last year.

Mr. Weymann drove a special Nieuport machine, which was the most speedy type of aeroplane built at that time, and was successful in landing and starting from a ploughed field, which many thought impossible for a very fast type of machine. It took the greatest skill to land such a speedy machine on rough ground, for he had to glide down with absolute accuracy, to land without a smash.

Among Army officers the keenest competition is developed, and it is only by a spirit of rivalry and a desire to excel that the best qualities in officers and men are brought out in times of peace. Of course in time of war there is a need which calls for the best there is in a man.

The needs of the Army and Navy aviators have developed some special features in machines built for their purposes. They want to be as far out in front of the machine as possible so they can have an unobstructed view, and so that if they should be so unfortunate as to be pitched out, they will be quite clear of everything. This is especially true of naval machines built to fly over the water. Military aeroplanes also should have a standard method of control, so that any Army or Navy aviator can operate any Army or Navy machine.

CHAPTER V

THE AEROPLANE AS APPLIED TO THE ARMY
(By Captain Paul W. Beck, U. S. A.) [1]

WHENEVER science discovers anything new or startling, such discovery is immediately tested by practical men of commercial or professional life to ascertain whether or not it can be applied to their business or profession. In civil life these tests are to determine whether or not this new discovery can be applied to cheapen production or benefit mankind in any other way. In the Army two tests are always applied: first, to determine whether or not the discovery can be used to kill the other fellow and, second, to determine whether or not it can be used to prevent the other fellow from killing us. These are the tests which have been applied to the aeroplane by the military. Let us see how these heavier-than-air machines have responded to these tests.

Can aeroplanes be used to kill the other fellow?

---

[1] In July, 1912, Captain Beck was granted by the War Department the title of "Military Aviator"; the first time that any American has been given this title, which implies finished skill in both aviation and military tactics, and for which all the army aviators are to qualify.—AUGUSTUS POST.

Our problem here is not ethical but practical; it is not based on the determinations of the Hague peace convention, but upon the actual capabilities of the machine from a physical standpoint, considered apart from humanitarian principles. In other words we do not discuss whether or not it is *ethically* right to use aeroplanes aggressively, but whether or not aeroplanes are *mechanically* capable of such use. The Army does not disturb itself with ethical questions until they become rules of International Law, and then it only considers them as being binding in their actual observance under the conditions imposed by such law. Meanwhile the Army, by preparation in time of peace, seeks to gain the fullest possible measure of information along the lines of investigation necessitated by the mechanical side of the question.

Considered from this standpoint, the question is repeated: can aeroplanes be used to kill the other fellow? Well, where may we expect to meet this other fellow? He will be armed, of course. He will be on the ground, on the water, or in the air. Wherever he may be we must get close enough to see him, while we must remain far enough away to keep him from having a decidedly better chance of hitting us than we have of hitting him. If he is on the ground or on the water we must fly over him. If he is in the air we must manœuvre our air craft so as to gain an advantageous position over him; one where we can shoot our machine guns or rifles while he is unable to use his similar weapons against us. That is where skill as an aviator and superiority of speed, climbing powers, and control of the machine will play a prominent part in deciding the supremacy of the air.

From the standpoint of the location of the enemy the problem can be reduced to two cases: one, when the enemy is on the ground or on the water, and the other when he is in the air. Against him in the first case we must use projectiles dropped from on high. These may be shrapnel, explosive shells or simply large, thinly encased masses of high explosive, depending on whether we are attacking individuals or animals in groups; gun emplacements, bridges, etc., or important strategical or tactical points such as arsenals, barracks, or parts of a defensive line.

Against the enemy in the skies we must use some small machine gun or rifle, in an endeavour to brush him aside and allow our own information-gathering aeroplanes to perform their functions unmolested.

But we are not progressing. Can aeroplanes be used to kill the other fellow? Well, assuming him to be located as we have assumed him to be, there are several other questions which must be answered before we can clinch the main issue. Can a man act as aviator and at the same time manipulate the mechanism that may be found necessary to the killing of the other fellow? If not, can an aeroplane be built that will carry at least two men, one as aviator and the other as manipulator of the death-dealing apparatus, and, at the same time, carry enough extra weight, i. e., fuel, to keep aloft long enough to accomplish the necessary flight and also carry the projectiles and dropping device? Yes. The two passengers may be estimated to weigh three hundred pounds. The dropping device may be estimated to weigh not to exceed fifty pounds. At least three known types of aeroplane carry six hundred and fifty pounds of weight for a continuous flight of two hundred miles in length. That leaves two hundred and fifty pounds that can be devoted to the carrying of projectiles.

So far the coast seems clear, but a small storm appears in the offing; can this two hundred and fifty pounds, or any considerable part of it, be dropped from a moving aeroplane without disturbing its equilibrium to such an extent as to render the machine unmanageable? Any weight can be dropped from the centre of lift without disturbing the equilibrium. Thirty-eight pounds have been dropped from one machine from a point three feet in front of the centre of lift without disturbing the equilibrium.

Admitting that the necessary weight can be carried and can be dropped, we next encounter the highly important question, what can we hit from a height of, say, three thousand five hundred feet? At this point the problem becomes one of pure fire control, and is directly analogous to target practice in our sea-coast defences. Since the aeroplane is moving forward at a definite rate of speed at the instant of dropping the projectile, it follows that there is an initial velocity given to the projectile. This velocity is dependent upon the forward speed of the machine and varies with it. Gravity exerts an influence on the drop of the projectile, which influence increases the speed of drop as the altitude from which the shell is dropped increases. The direction and force of the wind currents through which the projectile must fall are variable and they all exert influences tending to cause the projectile to swerve from its original course to a degree dependent upon their strength and the thickness of each stratum of air. The size of the target and, if it be animals or men, the direction and rate of movement of the target, are all factors to a successful hit.

Practice has shown us that the principal factors are the forward speed of the machine and the altitude. The variations due to wind currents through which the projectile must pass in falling are negligible. The only targets to be chosen will be sufficiently large and immobile to warrant an assumption that they can be hit. Aerial target practice will never degenerate to the sniping of individuals. It will be directed against ships, small boats, armies, cavalry, quartermaster and field artillery trains and similar large bodies of men or animals, or against the strategical and tactical points alluded to above.

The problem then simmers itself down to a more or less accurate solution of a method for determining the forward speed of the machine and its altitude, which, with a suitable set of tables and suitable mechanical devices for releasing the projectile at the proper instant, will produce a reasonably good target practice.

For some time the solution of the forward speed of an aeroplane seemed impracticable. It has now been solved by the simple use of a telescope, mounted on a gimbal so as to maintain its horizontal position and movable vertically along a graduated arc. By setting the telescope to read an angle of forty-five degrees and snapping a stop watch on an object which lies in the line of sight of the telescope—produced, and then swinging the telescope so as to point vertically downward, we can, by snapping the stop watch a second time as the sighting point again comes into the field of vision, ascertain the exact time it has taken the machine to cover the distance measured by forty-five degrees of arc. Our altitude is known by reading a barometer. We then have two known angles of a right triangle and one known side, viz., the altitude. By a set of tables, already made out, we can determine our forward speed.

Now, all of this is done as a preliminary to actually dropping the projectile. After we have the forward speed and the altitude we simply consult another set of previously prepared tables and read from those tables an angle. This angle shows the proper point of drop to hit another point on the ground somewhere in advance of the aeroplane. After picking the angle out of the table we set our telescope to read the known angle and, when the line of sight, produced, is on the objective, we release or "trip" the projectile. This has actually been done. Now I ask you the question, can an aeroplane be used to kill the other fellow?

Can an aeroplane be used to prevent the other fellow from killing us? Of course it is much superior to Santa Aña's mule for purposes of rapid departure from the scene of hostilities, but that is hardly the test we apply. It is, on the other hand, inferior as a shield to the ordinary breastworks constructed by armies in the field, but, again, that is not precisely the test to be applied.

The most effective way in which we can keep the other fellow from killing us is to find out where he is, what he is doing and how he proposes to accomplish his—to us reprehensible, to him laudable—object. Accordingly we apply the information test to the aeroplane. Can we use it to gather information of the enemy, his lines of communication, his lines of defences, his probable lines of advance or retreat, his rail and water communications, his artillery positions and gun emplacements, and a host of other things, all of which tend to produce success or failure in battle? In other words, can we use the aeroplane to prevent the enemy from killing us?

In order to make use of information there are two distinct steps which must be taken: First, it must be gathered; second, it must be communicated to the proper officers for transmission to the Commanding General in the field. No information is of value until it is communicated to an officer competent to act upon it.

This problem of information is then divided into two parts: the getting, and the transmitting. In getting information we must at once settle just how far the aeroplane will be available. There is a certain class of information, i. e., that concerning the road beds over which an army must move, the fords it must cross, the bridges it must travel over, the hills and valleys that might afford shelter for an offensive force or may be used defensively, the location, extent, thickness and amount of underbrush in woods, and much other, intimate, local knowledge that is of great and indispensable value to a commanding officer in the field. Such information can be gathered only from the ground. An aeroplane could be of use in such gathering only as a means for transporting the topographical sketchers quickly from point to point, allowing them sufficient time to do their work before again taking the air. Also an aeroplane would be of but little use in locating small bodies of the enemy.

Where the aeroplane would begin to be of use, however, is in the locating of the main body of the enemy, his defences, his artillery positions, in determining the outline of his position, the natural or artificial boundaries which cover and protect his flanks, his main arteries of supply, the strong and weak points of his line of defence, etc.

To accomplish these results the aeroplane must fly at a sufficient elevation to render difficult the hitting of a vital part of the machine or the aviator by hostile rifle or artillery fire. While the modern rifle in use in our army will fire a ball about three thousand five hundred yards straight in the air, it is generally accepted among aviators that an aeroplane would be practically safe, save from a chance shot, at three thousand five hundred feet. Of course there is a large chance that if enough rifles are directed at an aeroplane for a long enough time the machine or operator would be hit, at this altitude, but war is not a game of croquet, and the men who would man these machines in war would stand ready to take the risks demanded by the exigencies of the service.

The proper machine to act as a gatherer of information is one that can carry a pilot, passenger, and wireless outfit. It is proposed to equip all information-gathering machines with wireless and to this end a special set has been devised and is being tested out at the U. S. Army Signal Corps Aviation School. That the wireless will be a success there is no doubt, for certain simple experiments with crude apparatus have been already tried out with remarkable success.

I have said that military aviators propose to fly at about three thousand five hundred feet while seeking information. Perhaps this will be increased to about five thousand feet if it can be demonstrated that the reconnaissance officer can clearly discern, from that height, the points which are of military value. This officer will be aided by powerful field glasses, a camera and sketching case, and he will have at hand a wireless outfit which he can use in sending back whatever he may ascertain of value. Upon reporting back to the officer who sent him out he will turn over his sketches and photographs. It is thought that in this way very complete and valuable data will be available.

From an aeroplane or balloon the ground presents a very different appearance than it does from our usual man's eye view. It takes time and practice to determine just what the different strange-looking objects are, let alone to determine relative sizes and distances. On this account we

ARMY AVIATION 215

have concluded that the reconnaissance officer and pilot must both be trained at the same time. Since this is the case and since there is a decided mental and physical strain connected with long-continued flight, we have gone further and concluded that both officers who fly in the aeroplane must be pilots and both must be trained in reconnaissance duty. In this way each can relieve or "spell" the other.

There is much more to this than the mere acting as an aerial chauffeur. To be a successful military aviator a man must be an excellent cross-country flier. He must be an expert topographer or sketcher, he must understand photography and he must be a practical wireless operator, as well as have a knowledge of the theory of wireless. Above all, he must be trained in military art, that most elusive of all subjects. By that we mean that he must understand the military significance of what he sees, he must understand the powers, limitations, and functions of the three great arms—infantry, cavalry, and field artillery, whether used in combination or separately; he must know major and minor tactics to determine the worth or uselessness of a position; he must be able quickly and accurately to reduce his observations to a written report in order that the information gained may be of immediate use to his chief.

For all of these reasons we have concluded that

we must rely on commissioned officers of the regular army or organised militia, trained in time of peace to fulfil their functions in time of war. We can not place dependence on civilian aviators, for they have not had the training along the highly technical and specialised lines that are necessary. We can not rely on enlisted men of the army, for the same reason.

There is another class of fliers that will, undoubtedly, be of use in war time. These are the men to drive fast-flying, single-passenger machines for speedy messenger service between detached bodies of troops, or to drive the heavy ammunition or food-carrying aeroplanes to relieve a besieged place. These may well be chosen from the ranks of the civilian volunteers who would, without doubt, flock to our colours and standards at the whistle of a hostile bullet. There is plenty of room in war time for all of the aviators we can scrape together, be they civilian or military.

Two new types of aeroplane have been alluded to in the last, preceding paragraph; the fast-flying, quick-climbing racer and the slow-going, heavy-weight carrier. We are of opinion that there should be three types in all for military purposes. Of greatest importance and in greatest numbers we should have the middle-class machines; those capable of staying in the air for at least three hours of continuous flight, while carrying two men and one hundred and fifty pounds extra, of either wireless apparatus or machine gun and ammunition. Such a machine will climb two thousand feet in ten minutes, will travel above fifty miles an hour on the level, is perfectly easy to manage, and forms the back-bone of the aerial fleet.

One of these craft acting as a convoy, armed with a Benét-Mercier machine gun weighing about twenty pounds and with ample ammunition, could sweep the skies clean of hostile aeroplanes, while its mate, carrying reconnaissance apparatus and two officers, could gather the information which the Commanding General desires. The speed machine is for use as described above. The weight-carrying machine can carry about six thousand rounds of ammunition at a trip. Rifle cartridges weigh about one hundred pounds per twelve hundred rounds. This machine could carry enough emergency rations on one trip to subsist five hundred men for a day. It could make a speed of forty miles per hour with this weight and, in the course of a day, could, undoubtedly, make several trips of succour, provided the sending point were within fifty miles of the besieged place—which is the usual case.

And now, can an aeroplane be used to prevent the other fellow from killing us?

This is a very fascinating subject as a whole. The field opened is almost limitless; but the greatest idea of all is that through this conquest of the air we are approaching more nearly to that much longed-for era of universal peace. Through the aeroplane and dirigible, man is effacing artificial barriers; he is bringing the rich closer to the poor, the powerful closer to the weak. No longer can unwise and selfish potentates, be they royal, democratic, or financial, send forth their armies to fight while themselves resting safe and secure at home. The king in his palace or the money baron on his private yacht is in as much danger from these air craft as is the high private in the muddy trenches at the front. That touches the selfish side of things. At any rate, while the aeroplane will, probably, do more to promote peace than has any previous discovery, we of the Army are still busily engaged in finding out just what it will do in war.

CHAPTER VI

THE AEROPLANE FOR THE NAVY

(With an Account of the Training Camp at San Diego. By Lieutenant Theodore G. Ellyson, U. S. N.)

THE first active interest of the Navy Department in the practical side of aviation may be said to date from November, 1910, when Glenn H. Curtiss offered to instruct one officer in the care and operation of his type of aeroplane. Prior to this date the Department had carefully followed the development of the different types of aeroplanes, but had taken no steps toward having any one instructed in practical flying, as at that time there was no aeroplane considered suitable for naval purposes. Again, shortage of officers and lack of funds for carrying along such instruction were reasons for the delay in taking the initial step. There were unofficial rumours to the effect that there would be an aviation corps organised, and it was understood that requests for such duty would be considered, but it was looked upon as an event that would take place in the dim future. At this time Mr. Curtiss made his offer to instruct an officer at his flying field which was to be located in southern California, and, as it was understood that he had in view the development, during the winter, of a machine that could be operated from either the land or the water, his offer was immediately accepted by the Navy Department, and I was fortunate enough to be detailed for this duty.

The training camp was located on North Island, opposite San Diego, California, this spot having been selected on account of the prevailing good weather, and because there was both a good flying field for the instruction of beginners, and a sheltered arm of San Diego Bay, called The Spanish Bight, for carrying on the hydroaeroplane experiments. The camp was opened on January 17, 1911, and shortly thereafter seven pupils were on hand for training, three army officers, one naval officer and three civilians.

What was accomplished there is now history, namely the development of a machine that could rise from, or land on, either the land or the water, a feat that had never before been accomplished. It is true that one man had been able to rise from the water; but in attempting to land on the same he had wrecked his machine, so this could not be called a successful experiment. This same machine which had risen from the water and landed on the land and then risen from the land and landed on the water, was flown from the aviation field to the U. S. S. *Pennsylvania* by Mr. Curtiss, a landing made alongside and the aero-

NAVY AVIATION

plane hoisted on board with one of the regular boat cranes. No preparations had to be made except to fit a sling over the engine section of the aeroplane so that it could be hooked on the boat crane. The aeroplane was then hoisted over the side and flown back to the aviation field.

As I have said, the above paragraph is now history. What is not generally known is the hard work and the many disappointments encountered before the hydroaeroplane was a real success. Mr. Curtiss had two objects in view: First, the development of the hydroaeroplane, and secondly, the personal instruction of his pupils. The latter was accomplished early in the morning and late in the afternoon as these were the only times when the wind conditions were suitable, and the experimental work was carried on during the rest of the day, and, I think, Mr. Curtiss also worked the best part of the remainder of the time, as I well remember one important change that was made as the result of an idea that occurred to him while he was shaving. No less than fifty changes were made from the original idea, and those of us who did not then know Mr. Curtiss well, wondered that he did not give up in despair. Since that time we have learned that anything that he says he can do, he always accomplishes, as he always works the problem out in his mind before making any statement.

All of us who were learning to fly were also

interested in the construction of the machines, and when not running "Lizzy" (our practice machine) up and down the field, felt honoured at being allowed to help work on the experimental machine. You see it was not Curtiss, the genius and inventor, whom we knew. It was "G. H.," a comrade and chum, who made us feel that we were all working together, and that our ideas and advice were really of some value. It was never a case of "do this" or "do that," to his amateur or to his regular mechanics, but always, "What do you think of making this change?" He was always willing to listen to any argument but generally managed to convince you that his plan was the best. I could write volumes on Curtiss, the man, but fear that I am wandering from the subject in hand.

One of the results of the experiments at San Diego, was to show that such a hydroaeroplane, or a development of it, was thoroughly suitable for naval use. Although it was the first of May before Mr. Curtiss returned to his factory at Hammondsport, specifications, which were approximately as follows, were sent him and he was asked if he could make delivery by the first of July:—

"A hydroaeroplane, capable of rising from or landing on either the land or the water, capable of attaining a speed of at least fifty-five miles an hour, with a fuel supply for four hours' flight. To

NAVY AVIATION

carry two people and be so fitted that either person could control the machine.''

His reply was in the affirmative and the machine was delivered on time. Since that time this machine has been launched from a cable, which can easily be used aboard ship, and has been flown on an overwater nonstop flight, one hundred and forty-five miles in one hundred and forty-seven minutes. If such an advance has been made in a little over six months' time, what will the next year bring forth?

In my opinion the aeroplane will be used by the Navy solely for scouting purposes, and not as an offensive weapon as seems to be the popular impression. This impression is probably enhanced by the recent newspaper reports of the damage inflicted upon the Turks in Tripoli, by bombs dropped from Italian aeroplanes. Even could an explosive weighing as much as one thousand pounds be carried and suddenly dropped without upsetting the stability of the aeroplane, and were it possible to drop this on a ship from a height of three thousand feet, which is the lowest altitude that would ensure safety from the ship's gun fire, but little damage would be done. The modern battleship is subdivided into many separate water-tight compartments, and the worst that would be done would be to pierce one of these, and destroy those in that one compartment, without seriously crippling the gunfire or manœuvring qualities of the ship. In only one way do I see that the aeroplane can be used as an offensive weapon, and that is when on blockade duty, with the idea of capturing the port, ships out of range of the land batteries could send out machines with fire bombs and perhaps set fire to the port.

Innumerable instances could be cited where the use of an aeroplane for scouting purposes would have been invaluable. In recent times may be cited the blockade of Port Arthur during the Russo-Japanese War, and the blockade of Santiago, during the Spanish-American War.

Again suppose that several scouts were on the lookout for an enemy's fleet, and that they sighted the enemy's smoke. It has been proven that by modern scouting methods it is next to impossible for an enemy to start for any of several destinations, no matter how many miles apart, and not be discovered by the opponent's scouts before reaching their destination. The enemy's main strength, or battleships, will be covered by a screen, that is cruisers and torpedo boat destroyers, spread out many miles from the main body, whose duty it is to prevent our scouts from getting near enough to obtain any information. In order to obtain the necessary information our scouts would have to pierce this screen, and the chances are very great that they would be sunk in the attempt, or so crippled that they would be

ELLYSON LAUNCHES HYDRO FROM WIRE CABLE
(A) The start. (B) Leaving the wire

ROBINSON'S HYDRO FLIGHT DOWN THE MISSISSIPPI

unable to convey the information to the Commander-in-Chief. In any event, why run such a risk? If equipped with aeroplanes it would be an easy matter to send them out, and the information would be obtained in a much shorter time, without danger of the loss of a ship, and with the surety that the information would be secured. In this connection it must be remembered that there is nothing to obscure the vision at sea, that the range of vision from a height of three thousand feet is approximately forty miles, and that the wind conditions are always better than over land; that is, steady. These are simply a few instances of the value that an aeroplane may be to the Navy.

In my opinion, the ideal aeroplane for naval use should have the following characteristics: The greatest possible speed, while carrying two people and fuel supply for at least four hours' flight (not under sixty miles an hour speed, as this has already been accomplished), and, at the same time, capability of being easily handled in a thirty-mile wind. There are many machines for which this quality is claimed, but few that have really proved it. Double control so that either person can operate the machine. Ability to be launched from shipboard, without first lowering into the water, as on many occasions the wind at sea will be suitable for flying, whereas the sea will be too rough to rise from. Ability to land on rough water. The engine to be fitted with a self-starter. Also that the engine be muffled and the machine fitted with a sling for hoisting on board ship by means of a crane, and so constructed that it can be easily taken apart for stowage, and quickly assembled.

A search-light for making landings at night, and an efficient wireless apparatus, should also form part of the full equipment.

I did not make one of the requirements that the aeroplane be able to rise from the water, for in actual service it could always be launched from the ship. For practice work and for instructional purposes, it must be so fitted, but this could be a different rig if necessary. In the near future I predict that the aeroplane adopted for naval purposes will operate from a ship as a base and the great part of the instructional work will be done in the hydroaeroplane on account of the large factor of safety.

CHAPTER VII

GLIDING AND CYCLE-SAILING—A FUTURE SPORT FOR BOYS, THE AIRMEN OF TO-MORROW

(By Augustus Post.)

THERE is great popular interest in the problem of soaring, or flying as birds do, without any apparent effort, and also in gliding flights, or descending from a high altitude without the help of a motor.

Wonderful keenness of feeling on the part of an aviator, akin to that remarkable sensitiveness which is exhibited by all blind people, may be highly developed—for an aviator is just like a blind person in the air as far as concerns seeing the eddies, gusts, and currents, which are so dangerous to the balance of the machine—but the ability to advance and go ahead against the wind is as far off as the wireless transmission of power is to-day. It is necessary to have an up-current of air to enable a machine to soar and it is necessary to find where these upward blowing currents are. Any bicycle can coast down hill and a glider is only a coasting aeroplane, and it may be as difficult to find the right air current as to find a hill to coast down on a bicycle.

Great advances will be made in the art of aviation along the lines of training men in the art of handling an aeroplane. No opportunity is so good for this purpose as handling the machine as a glider with the motor shut off, or by practise with a regular gliding machine. Boys will naturally take to gliding, and as a glider was the first form of flying-machine and the easiest to build mechanically, there is every reason why sailing or soaring flights should be thoroughly mastered. The instinct which birds have which enables them to seek out and to utilise the rising currents of air in the wind and so to set and adjust their wings as to enable them to take advantage of these rising currents, is latent in the human mind and can be developed by practice to a point far exceeding that of birds, on account of man's superior intelligence. It is quite possible that some arrangement may be made by which an aviator can see the air and can prepare for or escape conditions that are not favourable to his manœuvres. It is clear that the wind gusts, swirls, and turbulences exist in the air, for they are quite evident when we watch a snowstorm and can see the snowflakes as they float, impelled now in one direction, now another, or as we see dry leaves carried about by a sudden gust of wind, or, even more clearly when over sandy plains we can see the great columns of dust ascending in the center of whirlwinds for hundreds of feet, carrying heavy particles to great heights. It is quite possible that birds can see the air itself by some arrangement of the lenses of their eyes which may either enable them to see the fine dust particles or to so polarise the light that the direction of its vibrations can be determined and the course of flight so changed that an air lane favourable to the path of the bird can be followed and by following out one stream lane among many, which has an upward trend sufficient to counteract the falling tendency, the bird can remain at an equal elevation.

Mr. Orville Wright has clearly demonstrated this to be possible by his experiments lately made at Kitty Hawk, N. C., where he was able to soar for ten minutes over the summit of a sand dune, so delicately adjusting the surfaces of his glider to the up-trend of the wind that he was falling or descending at the same speed that the wind was rising, and thus he seemed to stand still over one spot on the ground. After increasing his descent and approaching the ground, he was able by the delicacy of adjustment of his controls to change the relation in such a manner that the wind rising overbalanced the descending of the machine and he was carried backward and upward to the crest of the hill again, where he remained for a short time before again gliding downward to the level ground below. In the same manner that a boat sails against the wind by the force of the wind blowing against the sail, which is placed at an angle to it and which resists sidewise motion by the pressure of the water against the hull of the boat, a glider with horizontal sails set at the proper angle will also sail into the wind which blows against its surfaces and which makes the path of least resistance a motion forward and slightly descending with relation to the direction of the wind, but which, in the case of an upward moving current of air, may be a path rising in respect to the ground.

The development of skill in this art will come by practice, and young men will follow out the ideas and suggestions of the more experienced until we will have small, light, flexible machines with such sensitive control that, with small motors to enable them to rise or to get from one place to another, much as a bird flaps its wings when necessary to add a little to the power which it gets from the wind itself, or in rising from the ground, will be able to sail around and glide on the strength of the wind for hours at a time.

The clever aviator or real birdman with his keen instinct cultivated to a state of perfection, fitted with polarising glasses possibly, may seek out and utilise the various powers that are present in the air; adjusting his wings so that he will be supported by the upward motion of the air itself where it exists, or, by turning on his motor, moving from one rising column of air to another, upon which he may hover and circle around, steering clear of all those other air lanes which are leading in some other direction.

These glasses, by showing where the air waves are all of one direction, may reveal a current flowing in one way, while they may make great masses of air flowing in some other direction appear as of some other colour, say red, for instance; or, again, in another direction, all may look green, and it will only be necessary to keep where all is pure white.

Entirely new types of machines have been recently constructed in France called "aviettes" and "cycloplanes." These are machines like gliders which are mounted on bicycle wheels and small aeroplanes with wings which have aerial propellers turned by the pedals which drive them along the ground and through the air.

A contest was held in France in June, 1912, for a prize offered by the Puegeot Bicycle Company for the first machine of this type to fly a distance of about forty feet and later a second prize for the first machine to fly over two tapes one meter —three feet nine inches—apart and four inches high. Both of these prizes were competed for by machines without any motor and driven solely by man power. Over two hundred entries were received by the promoters of the contest, but no one accomplished the flight on that date of the public contest. Three days afterward, however, Gabriel

Paulhain succeeded in winning the prize put up for the second test. He flew eleven feet nine inches on his first trial and ten feet nine inches on the second, which was made in the reverse direction.

There seems to be great interest in this form of human flight, which was the original way of attacking the problem of flight itself. When the gasoline motor was perfected mechanical flight followed very quickly and was rapidly developed to a high degree of practicability. It is possible that with encouragement human flight may also become more common than it now is.

PART V

EVERY-DAY FLYING FOR PROFESSIONAL
AND AMATEUR

BY

GLENN H. CURTISS

WITH CHAPTERS BY
AUGUSTUS POST AND HUGH ROBINSON

CHAPTER I

TEACHING AVIATORS—HOW AN AVIATOR FLIES

TEACHING another man how to fly is a very important matter, in whatever way you look at it.

You can take a perfect machine and select ideal conditions and let everything be right for making a flight and then it is directly up to the pupil—he must do the operating of the machine, no one else can do it for him. In a single passenger machine, the instructor can clearly show how it is done and then the other fellow must do it. The trick in learning to fly is self-confidence and that must be gained by personal practise. Any man who wants to fly badly enough can fly.

Almost all of the aviators that have flown and are now flying Curtiss machines, like Hamilton, Mars, Ely, McCurdy, Beachey, and Willard and the army and navy aviators, have been practically self-taught although now we have a regular school under the supervision of Lieut. J. W. Mc-Claskey, U. S. M. C. (retired), who has had great success with his pupils. I have been flying for over four years and I feel that I don't know much about it yet.

The would-be aviator should go to a good school where the best facilities can be had and where there is a good large place to fly, without obstructions. The machine should be thoroughly mastered and every part understood. Training a man to fly does not, as I regard it, consist in putting him in an aeroplane and letting him go up before he knows how to get down again. Anybody may be able to go up in an aeroplane, but it requires skill and practice to come down without damage to man or machine.

HOW TO FLY

An aeroplane is supported in the air by its wings. These are placed at a slight angle to the direction in which it goes so that the front edge is slightly higher than the rear edge. This tends to push the air downward and the speed of the aeroplane must be great enough to skim over the air before it has a chance to flow away. You may have had the experience of skating over thin ice which would bend beneath your weight as long as you kept moving, although it would have broken if you remained in one place. This is precisely the same phenomenon, and as the water has not time to flow away underneath from the thin ice so the air is caught under the surfaces of the wings and the machine passes on gathering new air as it goes to support it, faster than the air can flow away. A curved surface is better than a flat one and to find just the proper curve to be most efficient at the speed at which the machine is to fly is a very difficult problem and must be determined by very careful laboratory experiments.

The various flying machines have different ways of accomplishing the control of the rudders for steering to the right or left, and up and down, for a flying machine is different from all other vehicles in this one respect. In addition to the steering, the machine must be balanced, and as the air is the most unstable of all mediums, how to maintain the equilibrium becomes perhaps the most important point in the construction of an aeroplane, as well as the most necessary one for the aviator to master. This is accomplished in various ways and is the characteristic feature of the different machines.

The Curtiss machine is considered one of the simplest of all. When it is remembered that Mr. C. F. Willard, my first pupil, learned to operate a machine with hardly any instruction it would seem that the mere learning to operate should not be a serious obstacle to overcome. If the air is still and there are no wind gusts to strike the machine sideways and upset it, flying is easy, but if the air comes in gusts and is rolling and turbulent even the best and most skilful operator is kept busy manœuvring the front rudder and endeavouring to keep the machine headed into the wind, and when it tips, moving the side controls to maintain the balance. With all of these movements it is no wonder that the aviator's mind must be active—there is no time to think, every movement and act must be absolutely accurate and the body must be under full control.

The operator sits on a small seat just in front of the lower main plane; directly in front of him is a wheel which he can push out or pull back. Pushing the wheel out turns the elevating surfaces so that the machine points down. On the other hand, pulling the wheel toward you points the machine up, causing it to rise higher into the air. Turning the wheel to the right or left steers the machine to the right or left in the same manner as a boat is steered by turning its rudder.

The operator now must consider how to balance the aeroplane. On each side at the extreme outer ends of the machine are placed small horizontal planes so hinged at their front edge that they may be turned up or down. They are connected together in such a manner that when one points up the other points down, thus acting as a "couple"; wires connect these stabilising planes to the movable back of the pilot's seat. This has a yoke which fits over the shoulders of the operator.

When the machine tips to the left the aviator naturally leans to the right or the highest side and the lever is moved to the right by the pressure

LEARNING TO FLY

of the shoulder. This causes the left hand stabilising plane to be pulled down so that it offers its surface at an angle to the wind and exerts a lift on its side while the right hand plane is turned the opposite way, which causes it to exert a depressing effect on its side; this tends to right the machine.

The operator must use his feet also for there is a pedal for the left foot which operates the throttle of the engine, causing it to go faster or slower, and one for the right foot which operates a brake on the front wheel, which helps to stop the aeroplane after it has landed and is running over the ground on its wheels.

THE FIRST STEPS

It is necessary to know every detail of the machine—every bolt, nut and screw, and the purpose each serves in the economy of the whole. It is absolutely essential for the successful aviator to know his motor. The motor is the heart of the aeroplane, and keeping it in good order is just as necessary to the aviator's safety as is the keeping of his own heart strong for any emergency that he may be called to face.

After becoming familiar with its workings, so that it becomes second nature to make the right movements, get into the machine and when the air is perfectly still run it over the ground.

When there is no more novelty in the sensation and the machine is in a good position to get up speed you raise the elevator a little and try making short jumps into the air. The other pupils standing in a group at the end of the field are usually hoping and praying that you will not smash the machine before their turn comes and so cause delay until it is repaired.

In San Diego, there was great rivalry between the Army and the Navy. Witmer and Ellyson used to get up by sunrise and go over to the island and take out the old machine we used for teaching, which was nicknamed "Lizzy." They did this secretly because there was only one machine and they did not want the Army to smash it and so keep them down on the ground. After making their practice, they would go home and come back later, pretending that it was their first appearance.

When the officers began their schooling they fell steadily into my way of looking at the problem, and not one of them spared himself bruised hands or grimy clothing. For the first ten days I did not offer them a chance even to give the motor its full power while they were in the aviator's seat. After they had worked around the aeroplane long enough, however, and were familiar with all its details, they were allowed to make "runs" over the half mile course, straight-away.

That is, they took their seats in the machine in turn, the propeller was started, and the machine propelled along the ground on its wheels, like an automobile, without being able to rise. To prevent the machine rising while one of the men was in it, the throttle of the engine was so arranged that it only got half power, which was not sufficient to give it lifting power, but enough to drive it along on the ground at twenty or twenty-five miles an hour. This "grass cutting," as the boys soon dubbed it, gave them the opportunity to become used to the speed and the "feel" of the machine. It also taught them to steer a straight course by using the rudder and the front control, and to practise balance by the use of the ailerons. After a few days of these runs the throttle was given full vent, allowing full speed on the wheels, but the propeller was changed to one without the usual pitch. Thus, while the engine would drive the aeroplane at full speed on its wheels, this propeller did not have enough thrust to lift it from the ground. In this way the military pupils got the advantage of the speed, acquired balance, and adjusted their control to suit it, without the danger of getting up in the air too soon.

A little later, when they had thoroughly accustomed themselves to these conditions, still another propeller was put on. This one had just sufficient pitch to lift the aeroplane from the ground, when well handled, and it would make "jumps" of from twenty to fifty feet at a height of a few inches or, perhaps, a few feet.

These jumps served still further to develop the ability of the men to control the machine and perfect their balance, and it gave them the first sensation of being in flight at high speed, though not high enough to do any great damage should one of them be so unlucky as to smash up. A smash-up was what we particularly wished to guard against at all times, not only because of the cost of repairs and the delay, but largely because an accident, even though it may do no injury to the aviator, may seriously effect his nerves. I have known of beginners who, while making rapid progress in learning to fly, suffered a complete setback just because of an unimportant accident to the machine in flight, or in landing. Eagerness to fly too soon is responsible for many of the accidents that befall beginners. An ambitious young man may become thoroughly convinced after a few jumps that all he needs for making a long and successful flight is the opportunity to get up a hundred feet or so. The first chance he has, he goes up as he had planned, and unless he is lucky or an exceptionally quick thinker, the odds are that he will smash up in getting back to earth again.

I have never seen any one more eager to fly, and to fly as quickly as possible, than were these officers. Probably they were following the military bent of their minds or, perhaps, it was the enthusiasm of the pioneer in a new science.

As a rule, the mornings at San Diego are fine. There is seldom any wind during the forenoon, except when one of the winter rain storms blows in from the ocean. We tried to get in as much work during this calm period as possible. The mornings were found to be the best for doing this work. It was most desirable, not to say necessary, that the pupils should have a minimum of wind during their early practice work. Even the lightest wind may sometimes give serious trouble to the beginner. A gust may lift the aeroplane suddenly and then just as suddenly die out, allowing the machine, should it be in flight, to drop as quickly as it rose. Such a moment is a critical one for an inexperienced man. He feels himself dropping and unless he keeps his head clear, he may come to grief through doing too much or too little to restore his equilibrium.

In the practice work all the officers, as well as two private students, C. C. Witmer of Chicago and R. H. St. Henry of San Francisco, used the same machine. This was one of the older types of biplane, with especially strong wheels, and with a four-cylinder engine. This type was selected as best adapted to the strain of heavy work. It had sufficient power, under its regular equipment, to fly well, but had not the very high speed of the latest type, fitted with eight-cylinder engines. For beginners, I consider the four-cylinder machines the best.

While most of the practice runs and jumps were made during the hours of the forenoon, when there was little or no wind, there was plenty of work on hand to fill in the afternoons as well. We were all the while experimenting with various devices, some of them new, others merely modifications of the old. All of these, whether new or old, involved many changes in the equipment of the aeroplanes. There was seldom a time when at least one or more of the four machines we kept on the island was not in the process of being taken down or set up. Besides, there was the long series of experiments with the hydroaeroplane, which were carried on from day to day without affecting the regular practice work.

These frequent changes in motor, propeller, planes, or controls, were always taken part in by the officers. Thus they became acquainted with everything about an aeroplane and knew the results produced by the changes. I consider this the most valuable part of their training.

All this "building up" process, as it may be called, that is, building up a thorough knowledge of the aeroplane until every detail is known, I believed to be necessary. I proceeded on the theory that confidence is sure only when the aviator has a thorough understanding of his machine, and confidence is the absolute essential to the man who takes a trip in an aeroplane. If the aviator has not the knowledge of what to do, or what his machine will do under certain conditions, he would better not trust himself in the air. Once the men learned to make the runs and jumps successfully and to handle the machine with ease and confidence, they were ready for the next stage of their training before they could be trusted to make a flight. This was to go as passengers. For the carrying of a passenger, I chose the hydroaeroplane.

This machine was not equipped with wheels for landing on the earth, when I first began to use it, but had all the equipment for starting from or landing on the water. We had built a hangar for storing it at night close down to the water on Spanish Bight, which gave us the smooth shallow water for launching it and hauling it out with ease.

First, the men were taken in turn as passengers for runs over the surface of the bay. On these runs I made no attempt to rise from the water. I wanted to give the men time to accustom themselves to the new sensation of skimming over the water at forty miles an hour, for that is the speed at which I was able to drive the hydroaeroplane. The machine would skim along under full power, with the edge of the float "skipping" the water as a boy skips a stone on a pond.

After this I undertook short flights, taking each officer in turn as a passenger, and keeping within fifty or a hundred feet of the water. At intervals I would make landings on the water, coming down until the float touched the surface, and then getting up again without shutting off the power. When these flights had been made for several days and the men had accustomed themselves thoroughly to the sensation of being in flight, I believed they had progressed far enough to be taken up for longer and higher flights over both land and sea. In these flights I used a machine equipped for landing on both land and water with equal safety.

One of the most important things that should be developed in the beginner, and, at the same time, the most difficult, is the sense of balance. Every one who has ever ridden a bicycle knows that the sense of balance comes only after considerable practice. Once a bicycle is under way the balance is comparatively easy, but in an aeroplane the balance changes with every gust of wind, and the aviator must learn to adjust himself to these changes automatically. Especially is a fine sense of balance necessary in making sharp turns.

Some aviators develop this sense of balance readily, while others acquire it only after long practice. It may be developed to a large extent by going up as a passenger with an experienced

LEARNING TO FLY 247

aviator. I have noticed that it always helps a beginner, therefore, to make as many trips as possible with some one else operating the aeroplane. In this way they soon gain confidence, become used to the surroundings, and are ready for flights on their own hook.

One by one the officers were taken up as passengers on sustained flights until they felt perfectly at ease while flying high and at great speed. The machine I used for passenger-carrying practice work was capable of flying fifty-five miles an hour without a passenger, and probably fifty miles an hour with a passenger. This speed gave the men an opportunity to feel the sensation of fast and high flying, an experience that sometimes shakes the nerves of the amateur.

All this took time. As I have said elsewhere, I did not want to force the knowledge of aviation upon the young officers. Rather, I wanted to let them absorb most of it, and to come by the thing naturally and with confidence. It was much better, as I regarded it, to take more time, and give more attention to the little details, than to sacrifice any of the essentials to a too-quick flight.

The men who had been detailed to learn to fly, I assumed, would be called upon to teach other officers of the Army and Navy and, therefore, they should be thoroughly qualified to act as instructors when they should have completed their work at San Diego. This is the view they took also, I believe, and I never saw men more anxious to learn to fly.

During the last period of instruction, when the men had gone through all the preliminaries; when they had learned how to take down and set up a Curtiss aeroplane; knew the motor, and how to operate it to the best advantage; in short, were thoroughly acquainted with every detail of the machine, they were ready for the advanced stage of the work. This was to take out a four-cylinder aeroplane for flights of from three to ten minutes' duration at various heights.

My instructions to all of the men were never to ascend to unaccustomed heights on these practice flights; that is, not to venture beyond the heights at which they felt perfectly at ease and capable of handling the machine, and to make a safe landing without danger to themselves or to the machine. These instructions were obeyed at all times. Perhaps the caution exercised at every stage of the instructional period had had its effect on the men and they felt no desire to take unnecessary chances.

When they were able to fly and to make safe landings in a four-cylinder machine, I considered that I had done all I could do to make aviators of them. I had tried not to neglect anything that would prove of benefit to them in their future work—things I had had to learn through long years of experiments and many failures. In other words, I tried to give them the benefit of all my experience in the many little details that go to make the successful aviator.

Given the proper foundation for any trade or profession, the intelligent man will work out his own development in his own way. I could only start the men along the road I believed to be the easiest and safest to travel; they had to choose their own way and time to reach the goal.

It has been a pleasure and satisfaction to work with the officers of the Army and Navy. Their desire to learn the problems of aviation, intelligently applied, has made the work easier than I had anticipated. The many little annoyances that often beset us are forgotten in the keen satisfaction of having been of some service to the men themselves, and above all to our War and Navy Departments.

A BULLETIN ISSUED AT THE CURTISS AVIATION CAMP

The course is divided into six parts or stages.

1st. Ground work with reduced power. To teach running in straight line.

2nd. Straightaway flights near the ground, just sufficient power to get off.

3rd. Straightaway flights off the ground at a distance of ten or fifteen feet to teach use of the rudder and ailerons.

4th. Right and left half circles and glides.

5th. Circles.

6th. Figure eights, altitude flights and landings without power and glides.

In the above stages of instruction the men should learn the following about flying:

FIRST STAGE

Learn to run straight, using rudder and keeping on the ground. The idea is to be able to control under reduced power. Student must be kept at this continuously until he is perfectly at home in the machine and accustomed to the noise of the motor and the jar and movement of the machine on the ground. This practice should be kept up from one to two weeks, depending upon the ability the student shows in handling the machine in this part of the instruction.

SECOND STAGE

Motor throttled, but with sufficient power to allow the student to jump the machine off of the ground for very short distances. Care must be taken in adjusting the throttle to allow for wind conditions, otherwise machine may be shot up into the air suddenly and the student lose control of it. Student should be also instructed during these jumps to pay attention to the ailerons to keep the machine balanced. The throttle can be gradually let out to full as soon as the student begins to acquire the use of the ailerons and keeps good balance.

THIRD STAGE

Student should be instructed to rise fifteen or twenty feet from the ground in straightaway flights, and use rudder slightly in order to become accustomed to its use and its effect on the machine in the air. As soon as the student has accomplished the above he may be permitted to rise to the approximate height of one hundred feet if the field is large enough and to glide down under reduced power. When he has done this successfully many times, let him repeat the above gliding with motor cut out completely.

FOURTH STAGE

Student may be permitted to rise to the height of twenty-five to fifty feet and make half circles across the field to the right and then to the left. These circles should be shortened or sharpened with increased banking on turns until they are sufficient for any ordinary condition or case of emergency.

FIFTH STAGE

The student may be permitted to rise to a height of not less than fifty feet, and if the field is sufficiently large, permitted to make long circles, gradually shortening these circles until the shortest circle required is reached. Student should be cautioned not to climb on the turns. He should be instructed to drop the machine on the turns, thus increasing the speed and lessening the possibility of slipping sidewise in banking. He should be instructed to land as nearly as possible on all three wheels at once. This may be accomplished by flying or gliding as close to the ground as possible and parallel to it, then slowing the engine and allowing the machine to settle to the ground.

SIXTH STAGE

In making figure eights for pilot's license, student should try to climb as much as possible on the straightaways between the turns and drop slightly on the turns. In making glides from high altitudes where motor is voluntarily cut off, it is best to start the gliding angle before the power is cut off. In case the motor should stop suddenly, the machine should be plunged instantly if machine is at sufficient altitude and considerably sharper than the gliding angle, in order to maintain the head-on speed, and then gradually brought back to the gliding angle.

A DAY AT HAMMONDSPORT—NOTE BY AUGUSTUS POST

The Curtiss Aviation Camp at Hammondsport broke all records on June 22, 1912, by the number of flights made in a day. In all, two hundred and forty flights were made. One hundred and twen-

LEARNING TO FLY

ty-six of these were with the practice machine called "Lizzie" and constituted straight flights for the length of the field and half circles. Sixty-four flights were made with the eight-cylinder practice machine, and consisted of half circles, circles, and figure eights. The other sixty flights were made with the hydroaeroplane.

The twelve students who made these flights, some of whom were taking the course in the hydro and land machine both, expressed themselves as pretty thoroughly tired out at the end of this strenuous day's work. One hundred or more flights are made practically every day in the week, but the twenty-second being a particularly fine day, this new record was made.

The day's flying used up a barrel of gasoline and four gallons of oil.—A. P.

CHAPTER II

AVIATION FOR AMATEURS

THE man who contemplates buying an aeroplane for his own use will be especially interested in three subjects: First, how difficult it is to learn to fly; second, how long it takes to learn; and third, what is the cost of up-keep. By difficult I do not mean dangerous; any one who has gone far enough to consider owning and operating a machine knows and discounts the element of danger, and as to cost, it is easy to get figures on the first cost of an aeroplane; what the investigator would like to know is what it is likely to cost him for maintenance, breakage, and so on.

With a competent teacher—and if ever competence was necessary it is here—learning to fly is neither difficult nor dangerous. Six weeks ought to be time enough to teach one to fly, provided the pupil knows something about motors and is apt in other ways. Contrary to popular belief, reckless daring is not one of the requirements for success. Indeed, a man who applies for a position as aviator with the announcement that he is a daredevil afraid of nothing under heaven, is very likely to be rejected for this very reason,

AVIATION FOR AMATEURS 255

and a pupil who has the common sense to know that there is no especial point in defying a quite impersonal force like gravitation will get up a much better start than one who has so little caution that he wants to get up in the air too soon. Caution is the great thing for the beginner. Let him learn the machine first from the ground and on the ground, learn the controls and find out what to do when he shall be up in the air. Then let him learn how it feels to run over the ground on the wheels. Then he will begin to make "jumps," little ones, then longer and longer, until he is free of any fear of the air. This comes sooner with some than with others, and it is said that in some rare cases fear of the air never exists at all, for the great aviator, the star performer, like any other great man, has to be born with certain qualifications and a good many of them. There is no reason, with the advancing improvement in the flying machine, why almost every one with a real desire to fly should not be able in a comparatively short time to learn to do so.

As for the third point, it will cost no more to keep an aeroplane than to own an automobile. The initial cost is the greatest. Of course, there are the same qualifications that obtain with the automobile—the cost of up-keep will depend upon whether you have many and serious breakages and whether the owner looks after his own machine. Should the owner prefer to hire a competent mechanic, his wages will be about the same as those of a first-class chauffeur. As for smash-ups, the expense of these would be considerable, but not as much as it would be if an automobile should have an accident. For contrary to the ideas of a good many of the uninitiated, it is quite possible to injure an aeroplane, and quite seriously, too, without in the least hurting the aviator. In this respect the hydroaeroplane is of course safest of all; I am reminded of a recent accident at Antibes, near Nice, France, where Mr. Hugh Robinson, who was demonstrating a Curtiss hydroaeroplane, suffered a badly wrecked machine without the least injury. Forced to make a quick landing, he chose, in order to avoid a flock of motor boats filled with spectators, to dive directly into the water. The shock threw him out of the machine and he swam about unconcernedly until a motor boat picked him up. Of course a similar sharp contact with the solid ground would have wrecked the aviator to some extent as well, but it is possible to put a hydroaeroplane completely out of commission, necessitating expensive repairs, and not be more than shaken up.

Really there is much less danger of smash-ups than the outsider would think, provided the aviator is a careful driver. The main thing is to have great judgment in choosing a time for flights. An inexperienced aviator should never take up his machine in an unsteady wind of greater velocity than ten miles an hour. The less wind the better, for the beginner. The dangerous wind is the puffy, gusty sort, and this should be avoided by any but the most experienced aviator. It must be remembered, however, that it is the variations and not the velocity of the wind which causes trouble.

Another item of expense to be taken into consideration is the transportation of an aeroplane from one place to another, for it does not always go on its own wings. This, however, is neither difficult nor expensive. I am able, for example, to take down my machines and pack them in specially constructed boxes so that they take up but a comparatively small space for shipment. The setting up process is not difficult, nor even complicated, and can be performed by any one having had the proper instructional term at a first-class aviation school. An illustration shows an aeroplane, in its case, carried on an automobile.

With regard to safety as a steady, every-day means of transportation, all of us, in and out of the profession, know that, as Mr. Hudson Maxim has said, to make the aeroplane a common vehicle for, say, the commuter, "It must be improved so that flights shall become more a function of the machine and less a function of the aviator." At present a great deal depends upon the man who is flying—especially upon his quick and accurate judgment and his power to execute his judgment instantly and automatically. The man who buys an aeroplane to fly knows this beforehand and takes it into account; indeed it is a question whether, if the flying machine were as safe as a rocking-chair, there would be so much fascination about it; but while the aviator will always have to take into account, no matter how the mechanism may be improved, a certain element of danger that must attend it, he may as well remember, to quote Mr. Maxim once more, that "the tenure of life of no automobilist is stronger than his steering gear."

It certainly is not looking too far ahead to forecast the entrance of the aeroplane into the commuter's life. The great mass of the people certainly will not take the air-line, any more than they are now coming in by automobile every morning, and yet how many business men—and not necessarily the richest—do make the trip, that twice a day they used to take in a railroad car, in the open air, with the exhilarating breezes of their own automobiles? Perhaps not these same business men, but a corresponding class, will undoubtedly reduce the dull hours of train travel by half and turn them into hours of delight by the popularisation of aeroplane transportation. As has been the case with every means of transportation that has shortened time of travel, the habitable zones around cities will grow larger and larger as places hitherto inaccessible open before

AVIATION FOR AMATEURS

the coming of the swiftest form of transportation known to man, and the only one not dependent upon the earth's surface, whether mountain, swamp, or river, to shape its course.

If we had a course only a few hundred feet wide from New York to St. Louis or Chicago, aeroplanes could go through every day and there would be little danger; indeed, even as things are now, it would be a much safer method of travel than by automobile, as well as of course much faster. Long lanes with grass on each side and an automobile highway in the middle would be of the greatest advantage to both forms of travel. In crossing mountains on the downhill side an aeroplane could glide for long distances at an angle of one to five, so that if the elevation were a mile high it could glide five miles before landing. And on the up-hill side it could of course land immediately and with ease.

To return to the amateur, it is always better to go around an object that you can not land on immediately. Landing is indeed one of the most important points for the amateur aviator to consider. If it is possible, watch all accidents and study them closely. I take every means I can to learn what causes an accident so as to guard against it myself. Strictly speaking almost everything about the art of aviation is being learned by experimentation and the causes of accidents, while not always exactly ascertainable, are of the greatest interest to builders and operators of flying machines, for out of the accidents of to-day often come the improvements of to-morrow.

While learning, and indeed whenever possible, you should examine the ground before attempting to fly over it. The pupil should inspect every inch of the course over which he is to fly, by walking carefully over it, noticing all the holes and obstructions in the ground. Then should it be necessary to land, for any cause whatever, he will know instinctively where to land and what to avoid in landing. Keep away from other aeroplanes, for the wind-wash in their wake may tip up your plane and cause serious trouble.

My advice to the amateur begins and ends with one injunction: "Go slow." Yes, for more than a month, "Go slow." It is hard to resist the temptation to try to do stunts; with a certain amount of familiarity with your machine, so that you feel you could do a great deal more than you are doing, and with some experienced and confident performer all but turning somersaults with his machine over your head, to the delight of the crowd, it is hard to resist giving one's self the thrill that comes from taking a risk and not being caught, but you will do the stunts all the better for going slow at first.

Mr. Charles Battell Loomis, the late American humourist, said once, in talking about the opening of the fields of air:

AVIATION FOR AMATEURS 261

"It was thought that the automobile was a machine of danger, but the aeroplane has made it comparatively safe. A man in an aeroplane a mile above the earth, taking his first lesson all by himself, is in a perilous position. He has not one chance in a thousand of ever owning another machine.

"A man who will fly over a city full of hard-working people is a selfish brute. Until a man is absolutely sure of himself he should always fly with a good-sized net suspended beneath his machine.

"The man in the street has always hated new things. He hated velocipedes, then bicycles, then safeties, then automobiles, then motorcycles, but he has not yet learned to hate the aeroplane. But wait until monkey wrenches begin to fall on Broadway or beginners begin to fall on the man in the street. Then he will be mad at the aeroplane—if there is anything left of him."

Allowing for the humorous exaggeration, there is this element of truth in this—that mechanical flight has as yet a strong element of uncertainty.

Yet there are certainly wonderful stunts to be done with a flying machine, and the fun is as much in the effect on the flier as on the audience; perhaps even more so. I would fly for the mere sport if I were not in the business, for there is a fascination about flying that it is unnecessary to explain and difficult to resist. You can chart currents of the sea, but the wind is such a capricious element that though there are, so to speak, outline maps that could be made of the general direction of the winds, there will always be a certain uncertainty about their conduct. Nevertheless there are so much greater possibilities in flying than in any other of the arts, that it is no wonder the amateur wants to develop them. And in conclusion I can say that an aeroplane in perfect condition is as safe as an automobile going at the same speed—*and I mean it!*

CHAPTER III

HOW IT FEELS TO FLY
(By Augustus Post.)

THERE is no one question that people ask more often than: "How does it feel to fly?" Perhaps a passenger feels more keenly the sensations of flight than an aviator because his mind is not taken up with the operation of the controls.

As for the passenger, he climbs into the flying machine, takes his seat beside the operator, and becomes at once the centre of interest to all the people standing by. If he is himself an aviator it is another matter, but if it is his first experience in the air, he is usually the object of a certain shuddering admiration, not unmixed with envy.

The motor is started, making a terrific noise that almost deafens him, and quite drowns the parting speeches and the efferts of the funny men present to improve the occasion. With perfect calm, without the least excitement, the aviator listens to the noise of the motor; he hears it run and carefully notes the regularity of the explosions. When all is ready, he waves his hand— the signal for the man holding the machine to let go. The machine runs along the ground, gathering speed, bounces a little, so that one hardly knows when it leaves the ground; the front control is raised, and the machine is in the air.

You feel the rushing of the wind, and things below seem dancing about down there. The machine keeps its exquisite poise in the air, sensitive to the slightest movement of the control. As it rises, the forward plane is turned a little down, and as the machine varies in its elevation, the plane is turned to bring it back to the level; it tips a little to one side and the aviator moves, as it were instinctively, to correct the balance. The rush of the wind by your face becomes more violent, and the machine pitches and balances as if it were suspended by a string or by some unseen force which holds it up in the air.

When the flight nears its end and the machine flies low over the aviation field, the fences and trees there seem in a moment to be rushing to meet one. The planes are pointed downwards, the machine descends, is caught up again by the control, and glides along level with the ground, skimming just above the grass. The wind moves it a little sidewise, perhaps, but the pilot, with the rudder, straightens the machine around until it points right into the wind's eye and the wheels are parallel with the direction of the machine over the ground. The control now causes the machine to come lower until the wheels strike the ground—it rolls along—bounces a little over the rough field

(A) AUGUSTUS POST FLYING AT THE FIRST HARVARD-BOSTON MEET
(B) AN AEROPLANE PACKED FOR SHIPMENT—POST DRIVING

CURTISS' PUPILS

(A) J. A. D. McCurdy racing against automobile, Daytona Beach. (B) Lieutenant T. G. Ellyson, U. S. N. (C) Mr. and Mrs. W. B. Atwater, pupils at San Diego

—the brake is set, and the machine comes to a stop.

The aviator jumps down, the passenger climbs out with somewhat less agility, perhaps, and expresses his very hearty thanks, the plane is turned around, the propeller started, and the machine flies off again, leaving the passenger to tramp slowly through the grass, contemplating the insignificance of the human creature who is forced to walk humbly along the ground. You may remember that the first time you descended from an automobile and began to walk, you seemed to yourself to be only marking time.

This new experience, though of the same nature as that, is far more impressive; not alone the difference in speed, but the whole character of the motion—the altitude, the rushing wind, the sense of something long awaited and now realised— sets the sensation of flight apart from any other, and makes him who once experiences it resolved to repeat the experience as soon and as often as possible.

The passenger is at once the object of eager inquiries as to how he felt, and he usually makes it his business to express his satisfaction whenever asked and sometimes without being asked, so there is little wonder that aviators are besieged by applicants for rides. A few months ago a lady who had been a passenger in an aeroplane was certain to get her picture in the papers; now there are so many that it would be difficult even to keep a record of them.

Now that we are coming to regard the aeroplane seriously, more from the practical and less from the grandstand side, it may be noted without fear of loss to gate receipts, that its dangers have been greatly exaggerated. Rational flight is hardly any more hazardous than motor speeding, steeple chasing, and many other sports, not to mention football! Engines stop and planes split, but steering gear breaks and horses stumble. Danger lurks everywhere, but we disregard it because the chances are long in our favour.

The real danger in aviation lies in the *chances* men take as *desire* lays hold upon them; chances the dangers of which they fully realise, but disregard for various causes. There are so-called "holes in the air," but they are hardly more numerous than gullies in the road. High wind is dangerous, but the aviator can often avoid its perils if he will. Briefly, aviation confined to its now well-defined limitations, is a thoroughly rational sport.

The "queer" sensation of flight comes in a quick rise, dip or short turn, and you can experience the same sensation in the elevator of a New York sky-scraper, Ferris wheel, shoot-the-chutes or even the back yard swing, for that matter! Dizziness from height is not experienced, for one sees the landscape spread out from high up and afar off, as if from a sheltered balcony; the tendency is not to look down but away.

While the rush of air is tremendous, it is not disagreeable, and one even forgets the deafening, unmuffled motor in the indescribable joys, mainly because of the wondrous charm and variety of the landscape which we have known only in *detail,* ignorant of its beauty as a *mass.* Apprehension, shuddering, gruesome, childish apprehension perhaps, at the starting, replaced by profound security as mastery, perfect mastery, is apparent; a sense of joyous freedom following as the marvellous world below is revealed. Like an exquisite monotone in low relief it is, each note of colour with its value and in perfect harmony with the whole; ever subtly changing, always some new surprise, some unexpected revelation, lifting one on the wings of exaltation.

The popular literary vehicle of to-day, rivalling the "fairy coach of Cinderella," is without question the alluring aeroplane, fitted with all the latest improvements: tachometer, inclinometer, animometer, barograph, aneroid, compass with map holders, lights, and all the modern conveniences and aviation equipment, including a wireless telegraph outfit, having shock absorbers for landing and an enclosed limousine cabin with mica or celluloid windows, in which not only can our spirits be wafted about, but in which we may enjoy all the material comforts of speedy travel, free from present annoyances and inconveniences, and without requiring the inflated rubber suits which Mr. Rudyard Kipling so kindly provided for his passengers on board the now famous "Night Mail." Vehicles of this description already exist and an "aero-bus" has carried as many as thirteen passengers besides its driver. It is confidently predicted that twenty passengers will soon be carried in an aeroplane at one time.

There is no doubt but that in flying the higher faculties are called into play. No such elaborate preparation is necessary for learning to drive an automobile, but some instruction is usually found necessary when learning how to balance a bicycle for the first time and until confidence is secured, as is also the case in learning to swim. A good chauffeur does not necessarily make a good aviator even though he have exceptional ability as a driver of racing automobiles, although I think that an aviator might make a good driver of a racing automobile. This seems to indicate clearly to my mind that there is some additional quality required in flying. I know of one case where a successful automobile builder and driver killed himself on account of desperation over the fact that he could not master flying.

Actors and men with a keen sense of feeling seem to do well in the air. They seem to get the "feel of the air," or to have the delicate sense of touch which is required to handle an aeroplane among the illusive vagaries of the atmosphere, and to be able to sense its rapid action and feel its ever-changing conditions almost before they take effect. One must be absolutely *en rapport* with his machine, as an expert horseman is part of his horse or his horse is part of him; such a rider stands out from all the rest, a beautiful sight to see and an expression of the poetry of motion; such also is the manner of the master at the piano, whose very soul is in tune and vibrating with every subtle and rich harmony of the instrument, feeling at the same time the ever-changing mood of his audience as he sways them or is swayed by them in turn, keeping in close sympathy with their thoughts as well as suggesting to their minds the trend that they shall take.

AVIATING AND BALLOONING

The sensations which an aviator has during great flights of both duration and altitude are somewhat comparable to those of the balloon pilot [1] who sails in the sky far above the earth,

[1] Mr. Post is not only intimately connected with the development of the aeroplane but also one of the most capable practical balloon-pilots in the world. Mr. Post accompanied Mr. Allan R. Hawley in October, 1910, when the balloon "America II," representing the United States, broke the world's competition record and won the Gordon Bennett balloon cup by sailing one thousand one hundred seventy-two miles—from St. Louis to Lake Tschotogama, in the wilds of Quebec. The trip took forty-six hours. This record still stands as American distance record. Mr. Post

feeling a peculiar realisation of the immediate presence of the Supreme Being, overwhelmed with the magnitude of the universe, with a sense of being a part of it, untrammelled, unaffected by ordinary things, surrounded with extraordinary conditions, supersensitive and yet keenly realising, now, matters of vast importance; now, minutely weighing his life in his hands as if it were something far removed from himself; breathing an air full of vigour and inspiration, with a sense of exaltation pervading every cell of the body—is it a wonder that men enjoy such delights and really live only when they can cast off mere existence and rise either to the contemplation of such experiences by reading and thinking about them or to a full realisation of these experiences by actually trying them out personally? Such moments, rapidly passing—moments each going to make up our individual life—are usually but too few.

Is it then a wonder, that, after actual days of such vivid living, upon descending to earth or coming back among people, one should look at those who gather around about one as some kind of lower order of animal, that it should take a few moments to feel their presence gradually dawning upon him, and to bring his faculties slowly back where they can begin to understand

also holds, with Mr. Clifford B. Harmon, the American endurance record of forty-eight hours, twenty-six minutes.—THE PUBLISHERS.

what these bystanders are thinking and talking about?

This seems but a dream, but is in reality an actual experience of a return to earth after two days spent in the air and a visit to regions over four miles above its surface, much of the time out of sight of this dear old sphere, when ears had become unaccustomed to sound, and so impaired by the change of pressure due to the high altitude that we could not, for some time after landing, hear when spoken to. Our own voices rang hollow and stuck in our throats, and our thought had become unattuned to those expressed by the gaping, wondering crowd, struck dumb at the sight of our arrival, and standing like cows in the pasture when you walk among them.

Such is the state of mind in store for the airman, the artist, the thinker, the person desiring to become isolated for a while—to feel as Adam felt in all reality, when he stood in the midst of the garden of Eden, monarch of all he surveyed. This appeals strangely to the imagination but when it becomes a reality by virtue of actual experience, it also becomes a sensation most difficult to express; for so few people understand what you are talking about, few having had the sensations of being removed from this world and coming back again to it.

CHAPTER IV

OPERATING A HYDROAEROPLANE
(By Hugh Robinson.)

THE general impression among aviators and manufacturers of aeroplanes is that the hydroaeroplane is rapidly becoming the flying craft of the future, by reason of its ease of control, extensive bodies of water upon which to operate it, and, above all, its safety.

It is practically impossible for the operator of a hydroaeroplane to suffer injury in case of accident. Even in the worst kind of an accident, the most that can happen to the operator is an exhilarating plunge into salt or fresh water as the case may be, with the beneficial effects of a good swim if so desired, otherwise, the operator may "stand by" the wreckage, which cannot possibly sink. The several pontoons, together with the necessary woodwork to construct the planes, etc., furnish ample buoyancy to support the machine and operator even in case of a total wreck, which rarely ever happens. One can bang down upon the water with a hydro in any old fashion, and beyond a tremendous splash nothing serious happens.

OPERATING A HYDRO

Of course, this article refers entirely to the Curtiss hydroaeroplane, which I have been operating since its invention. The Curtiss pontoon is divided into six water-tight compartments, three of which will support the machine under average conditions. Recently, while the writer was abroad, a demonstration was made of these compartments for safety in case of accident to any part of the pontoon.

This demonstration took place at Monaco, and consisted in removing the drain plugs from two compartments, after which the hydro with pilot and passenger was pushed out into the harbour and allowed to stand thirty minutes to let the opened compartments fill with water, after which the motor was started and a flight made without the slightest difficulty.

The operation of a hydro is very similar to that of the ordinary land machine—only, if anything, considerably easier and more simple. The start of the hydro is simply starting the motor while the hydro is resting on the land or bank of the lake or river, with the front towards the water. The operator takes his place, and on opening the throttle gradually the thrust of the motor slides the apparatus along the ground, or planks if ground be unsuitable, and into the water. The pontoons being fitted underneath with steel shod runners makes it possible to start on rocks, gravel, or in fact most any reasonable surface. The finish can be made in the same manner, without assistance.

It is possible to start the hydro on dry land if the surface is reasonably smooth, with the assistance of one or two mechanics. It is also possible, in an emergency, even to land on the earth with the hydro pontoon attachment; and, of course, with wheels attached to the landing gear, one can come down on land as with the ordinary type of machine.

Once out upon the water, the operator rapidly increases his speed by opening the throttle, taking care, however, to accelerate gradually, to allow the pontoon to mount the surface of the water without throwing an unnecessary amount of water into the propeller. Once a speed of twenty-five to thirty miles an hour is obtained, the pontoon skims lightly over the surface of the water. As the ailerons do not become effective until the machine acquires considerable speed, the small floats on the lower ends of wings maintain the balance until necessary speed is acquired. The small flexible wooden paddles on the lower rear ends of the wing tanks slide over the water and exert a great lifting effect, thus rigidly preserving the balance on the water at slow speeds or standing, and also preventing damage to wings in case a bad landing is made whereby one wing strikes the water first. In such a case, instead of the wing digging into the water, the paddles

OPERATING A HYDRO

cause a glancing blow which levels the machine automatically.

When the machine has acquired a certain speed it leaves the water in exactly the same manner as on the land and immediately increases its speed, due to the released friction from the water. It also has a slight tendency to jump into the air due to the released friction between the boat and water. Once into the air, the operator is the same as with the regular land-equipped Curtiss aeroplanes.

The landing is made in the ordinary manner, bearing in mind to keep the boat as near level fore and aft as possible, and if the water be very rough to allow the tail of the machine to settle on the water first. This will prevent any possibility of sticking the front of the boat into an unexpected wave.

As should be the case with any aeroplane, it is advisable to start and land against the wind if there be much, but this is not compulsory. The hydro may be landed even while drifting sideways, in an emergency case. It is obvious that to do this with a land machine would be to invite disaster.

The writer saw a forcible demonstration of the one and two pontoon types of hydros during the Hydroaeroplane Meet in France, and he had the only machine there with the single pontoon, and also the only one able to go out on rough water. He successfully made flights and landings in waves six to eight feet high, whereas three hydros of the two pontoon type were wrecked in waves less than two feet high. The single pontoon-equipped hydro may be dragged out on the banks any place where a space two feet wide may be obtained, and on my recent trip down the Mississippi, I had occasion to rejoice in this fact and put it to a practical test, as I was hauled out on shores between large rocks or stumps in several instances. The turning of the hydro is accomplished by simply turning the rudder and leaning towards the turn, the same as on a bicycle, allowing the motor to run on reduced or half throttle.

The exhilaration of flying a hydro cannot be described on paper. It is the fastest motor boat in the world, and to be able to approach a launch and jump over it and observe the consternation of the passengers is the keenest pleasure imaginable.

The hydro may be used solely as a motor boat if desired, at a speed of sixty miles per hour, without a drop of water ever touching its passengers, or if weather be favorable, flights may be made at will of the operator.

The surface of a river or lake offers the ideal condition for landing or starting an aeroplane, and these are more numerous than suitable grounds for land machines, besides this the air conditions over water are always better than over land, due to its unbroken surface, which does not obstruct the air currents as do trees, houses, etc., on land.

An automatic safeguard exists in the hydro to prevent accidents, such as has caused the loss of lives on land, and that is as follows:

It is possible to rise in an ordinary land machine with too little power to make a turn or climb fast, and as a result get a bad fall. Owing to the fact that there is a suction between the water and the pontoon it requires more power actually to leave the water than to fly once the plane is in the air. This fact prevents a hydro taking flight with too little reserve flying ability, and once in the air the operator may be sure of a considerable reserve of power to enable him to fly strongly and safely under all conditions.

PART VI

THE CURTISS PUPILS AND A DESCRIPTION OF THE CURTISS AEROPLANE AND MOTOR

BY
AUGUSTUS POST

CHAPTER I

PUPILS

ALL great masters have been represented by pupils who have done honour to their teacher and have achieved personal success in a large measure. Mr. Curtiss is no exception to this rule, for he has taught more than a hundred pupils.

There have been representatives of all classes and all nationalities. The list includes all trades and professions, from horse trainers to bankers. And in all these have been pupils from thirteen nationalities including Russians, Germans, French, Canadians, Scotch, Irish, English, Japanese, Indians, Cubans, Mexican, Spaniards, and Greeks.

Instruction has been given in all languages, including the sign language. Some nationalities are naturally a little harder than others to instruct, largely because of national characteristics of thought, and also for the reason that in a southern climate those native to it are often unaccustomed to the rapid action necessary at times in flying.

Negroes have not yet as a class taken to aviation, but there is one Chinaman in California, Tom Gun, who has been successful as an aviator. But conspicuous among the list of pupils is the number of Army and Navy officers of our own, as well as of foreign countries, that have graduated from the Curtiss School.

Hydroaeroplane operation has also been taught to a number of pupils both at Hammondsport, N. Y., and at San Diego, California, where the training camps are located.

The life that the pupils lead at these schools is most interesting and healthful. The students get up early, sometimes at four in the morning, when it is just light enough to see and when the air is usually calm and the best conditions for learning to fly exist. Pupils are outdoors practically all day, flying, or working on the machines when any thing breaks or goes wrong. Many pupils have engaged in exhibition flying after completing their course of instruction, and among the large number of very excellent aviators that have followed in Mr. Curtiss' wing beats (for you can hardly say foot steps) have been some of the foremost aviators in the world and men whose fame and exploits are household words to-day.

A partial list of some of these men at present active in the field is here given:

Chas. F. Willard, Hugh Robinson, Chas. K. Hamilton, J. C. Mars, C. C. Witmer, R. C. St. Henry, Lincoln Beachey, Beckwith Havens,

CURTISS' PUPILS

Beckwith Havens
Chas. K. Hamilton

C. C. Witmer
J. A. D. McCurdy
Chas. F. Walsh

Cromwell Dixon
Chas. F. Willard

Lieut. T. G. Ellyson, U. S. N.; Capt. P. W. Beck, U. S. A.; Lieut. J. H. Towers, U. S. N.; William Hoff, J. B. McCalley, S. C. Lewis, C. W. Shoemaker, W. B. Atwater, Al. Mayo, Al. J. Engle, J. Lansing Callan, G. E. Underwood, Irah D. Spaulding, C. F. Walsh, Carl T. Sjolander, Fred Hoover, E. C. Malick, Ripley Bowman, T. T. Maroney, C. A. Berlin, H. Park, W. M. Stark, R. H. McMillan, F. J. Terrill, Francis Wildman, F. J. Southard, Lieut. P. A. Dumford, W. B. Hemstrought, Earl Sandt, R. B. Russell, Lieut. J. E. McClaskey, W. W. Vaughn, Barney Moran, M. Kondo, J. G. Kaminski, Mohan Singh, K. Takeishi.

Among those in this list who have done wonderful things, it might be interesting to mention some of the marvellous feats of daring as well as a few of the achievements of Lincoln Beachey, who is credited with being the greatest exhibition aviator in the world.

At the meet in Chicago in the summer of 1911, Beachey flew more miles than any other aviator. He flew all the time and was in the air during all the flying hours in one contest or another. He did all the special tricks in the air that were known, he carried passengers, won speed races, and established a new world's altitude record at 11,642 feet. After flying as high as he could, at Chicago, with a seven gallon tank full of gasoline, Beachey came down and said: "To-morrow I'll go higher." He had a ten gallon tank fitted to his machine, filled it full up to the top, and started right up from where his machine was standing on the ground, so as not to waste a drop of gasoline, and flew up and up until it was completely exhausted and his motor thus compelled to stop, but not until he had set the world's record at 11,642 feet. He deliberately started out on this trip to climb up as long as his fuel would last. He knew his motor would stop and he would have to glide down. It was not an unintended glide but it was the longest glide on record. He brought out all the points and possibilities of his machine; distance, speed, weight-carrying, and altitude. Wilbur Wright said: "Beachey is the most wonderful flyer I ever saw and the greatest aviator of all." Calbraith P. Rodgers said upon his arrival at Los Angeles after flying across the American continent, a distance of over four thousand miles, "Beachey's daring flight down the gorge of Niagara and through the spray of the falls was a greater achievement than mine." Beachey has been remarkably free from serious accidents even though now he pitches straight down from the sky, seeming to fall straight to the earth and just catching his machine up in time to avoid striking the earth.

At Hammondsport on July 29th, 1912, Beachey was trying out a new model military type and he ascended six thousand five hundred feet in fifteen minutes, while he came down in one minute, making one of his perpendicular dives with the engine still. The whistling of the wind through the taut wires of the machine could be heard half a mile away. On this occasion one of the lady visitors to the testing grounds, who had never seen Beachey fly before, thinking that he was falling and would surely strike the ground and be dashed to pieces, fainted. Beachey said, "Flying did not come to me at first but it seemed to come all of a sudden and then it came big."[1]

Once Beachey had to land in a very small place surrounded with trees and the only way he could do it with the fast machine that he was driving was to kill its speed in the air by skimming over the trees, shutting off his motor, and gliding along to the place where he wanted to stop, and then

[1] Ralph Johnstone said in a conversation about experiences while learning to fly, "I learned to fly all right but one day when I was up in the air pretty high I seemed to forget all about it and how to operate the controls. I tried them and tested how they worked and it seemed to me that I learned all over again, but it did seem funny to me for just a few minutes." Geo. W. Beatty said, "When I was flying at Chicago, in the contest for duration, when the weather was calm, and I had nothing else to do but sit and think while the machine flew on, round and round, lap after lap, I would look out at a wire and watch it as it vibrated and wonder if it was possible for it to break, while I realised that I could not get out to fix it. This worried me more than flying in a high wind. It seems more natural for me to fly than not to. I have been in the air on an average of two hours every day for over a year."

pointing the machine up suddenly, very much as a bird comes to a stop, and then "pancaking" down, as it is called when you come down "ker-flop" like a pancake.

Beachey broke a wheel by this performance and he has worried over that little breakage as much as another man would over smashing up a whole machine.

Beachey flew from New York to Philadelphia in company with Eugene Ely and Hugh Robinson in August, 1911, winning the first inter-city race to be held in the United States.

Among the skilled operators of hydroaeroplanes is Mr. Hugh Robinson who flew down the Mississippi River in the spring of 1912, carrying mail and covering the river course between Minneapolis, Minn., and Rock Island, Ill. Mr. Robinson also went to France in May of 1912, and competed in the first contests and races ever held in this new sport at Monte Carlo. Since his return to America, Mr. Robinson has been the instructor in hydroaeroplaning at Hammondsport.

CHAPTER II

A DESCRIPTION OF THE CURTISS BIPLANE

NO type of aeroplane is more familiar in America than the Curtiss biplane. By long experimentation, this machine has been developed for practical use; and is now used for military purposes in Russia, Japan, Italy, Germany, France, and the United States. The machine is of the general type known as "biplane," in which there are two sets of wings, or surfaces, one being directly above the other. This type of machine seems to be the most favoured by Americans, for it not only allows of a greater spread of lifting surface for a given width of plane than in the monoplane, or single-wing type, but also it is much stronger than other machines of the same weight, as its design permits of a system of bridge-trussing known as the "Pratt Truss." In the Curtiss machine this feature is especially pronounced, because of the greater safety which rigid planes have when compared with the flexible wings.

The woodwork of these aeroplanes is entirely of selected spruce and ash, all the posts, beams, and ribs being laminated. The propeller is a particularly difficult piece of laminated work, being built up of from twelve to eighteen layers of thinly cut wood, while the upright posts of the central section are made up of ash and spruce, the heavier and more flexible wood forming the core. A feature of strength is to be found in the double trussing which is placed in all of the vital parts of the aeroplane, where the greatest strength is required. All this trussing is made with a cable of galvanised steel wire tested to withstand a pulling strain of nearly half a ton.

Transportation and military use have been especially considered in the construction of the planes. The upper and lower planes are made up of interchangeable panels, which are so joined together that the machine is easily assembled and taken apart and may be transported compactly in two flat boxes which scarcely make one full wagon load, as indicated in an illustration in this book.

The wing-panels are made up with a light and strong wooden framework covered with cloth especially made and treated with a rubber coating for the purpose. The curved ribs are laminated also and the panels held together by a system of trussing which gives them great strength. These panels are covered both top and bottom.

Light and strong bamboo rods extend to the front of the main planes, supporting the elevator or forward horizontal surface, which acts as a rudder to steer upward and downward. Similar bamboo rods at the rear support the vertical rudder and rear elevators and stabilising plane. Front and rear elevators work in conjunction with each other so that as the front of the machine is directed up, the rear of the machine is depressed by the two rear elevators, called "flippers" from their resemblance to these appendages of a seal or a turtle, each of which is controlled by an individual set of cables, so that if one should break or get out of order the other may be used independently. The front or rear elevators are sufficient to maintain the fore and aft balance of the machine in flight, so if anything happens to one the other will enable a safe landing to be made. Some aviators take off the front elevating plane entirely, relying solely upon the two rear ones for horizontal control.

The elevators and the vertical rudder are manipulated by a single steering post at the top of which is the steering wheel. Turning the wheel to the right or left steers the aeroplane to the left or to the right as a boat or an automobile is steered, while pushing the wheel forward directs the machine downward and pulling the wheel causes it to rise, a system of control in accord with the natural impulse of the operator.

To maintain the lateral balance of the aeroplane, there are small movable planes, or "ailerons," attached at the ends of the main framework, midway between the upper and lower planes, at the rear. These ailerons are so arranged that the front edge remains in the same position; while one swings upward, the other swings downward, at the back, thus giving an upward pressure of air on the under side of the one, while the other is depressed by the air which strikes it on top. This movement is controlled by a movable back to the aviator's seat or a frame or yoke which fits around the shoulders of the aviator in such a way that he moves the ailerons to the proper position when he leans to the high side of the aeroplane as it tilts and is thus able automatically to correct its balance.

The motors with which the military and cross-country models are equipped are of the eight-cylinder "V-shaped" type, developing sixty and eighty horse-power. The propeller is attached directly to the motor shaft, thus doing away with any necessity of gearing, which consumes power, increases the risk of breakage, and decreases reliability. The speed of the motor is controlled by a throttle opened and closed by a movement of the left foot.

The seat for the aviator is placed well forward of the main planes, giving him a clear view not only ahead, but also straight downward. On the military model, a passenger-seat is provided immediately beside that of the aviator, and a dual system of control makes it possible for either pas-

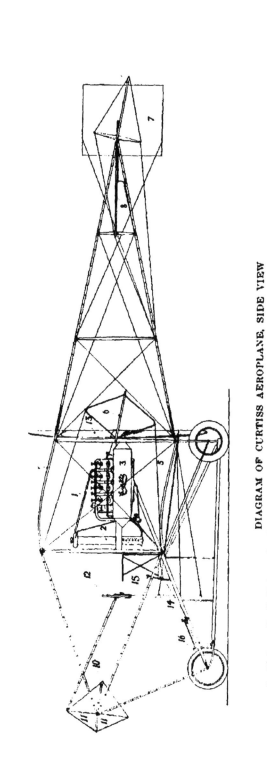

DIAGRAM OF CURTISS AEROPLANE, SIDE VIEW

1, Motor; 2, Radiator; 3, Fuel Tank; 4, Upper Main Plane; 5, Lower Main Plane; 6, Aileron; 7, Vertical Rudder; 8, Tail Surface; 9, Horizontal Rudder, or Rear Elevator; 10, Front Elevator; 11, Vertical Fin; 12, Steering Wheel; 13, Propeller; 14, Foot Throttle Lever; 15, Hand Throttle Lever; 16, Foot Brake.

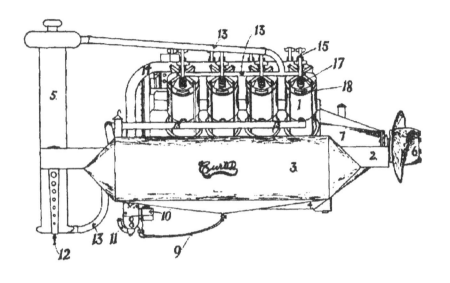

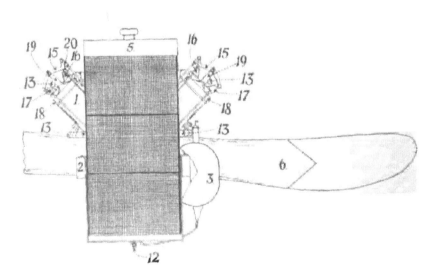

DIAGRAM OF CURTISS MOTOR, SIDE AND FRONT VIEWS

1, Cylinder; 2, Engine Bed; 3, Fuel Tank; 4, Oil Pan; 5, Radiator; 6, Propeller; 7, Crank Case; 8, Carbureter; 9, Gasoline Pipe; 10, Air Intake; 11, Auxiliary Air-pipe; 12, Drain Cock; 13, Water Cooling System; 14, Gas Intake Pipe; 15, Rocker Arm; 16, Spring on Intake Valve; 17, Spring on Exhaust Valve; 18, Exhaust Port; 19, Rocker Arm Post; 20, Push Rod.

BIPLANE PARTS 291

senger to operate the machine independently of the other.

The aeroplane is mounted upon a three-wheeled chassis with one skid extending from front to rear, the whole landing gear being built strong and rigid to withstand the shock of landing, the most dangerous part of flying.

Elaborate tests are made of the different parts of the machine; the panels forming the surfaces are tested by loading them with gravel until they break and weighing the amount of gravel heaped upon them before they give way. These tests have shown a factor of safety in excess of any strain that could be put on the machine in the air.

The strain on the various wires and cables is also measured, with a special instrument made for that purpose, as seen in an illustration. Every conceivable test has been tried which could give information that would lead to any improvement in strength to withstand strains, in addition to the complete knowledge that has come from actual tests under all conditions in the air, and on the ground itself, by expert flyers who have done almost everything that it is possible to do with the machine as far as trying to find its weak point is concerned. Dives almost straight down with abrupt turns at the end of the drop put many times the ordinary strain on every part. Rough landings also show up any lack of strength or fault in the design of the running gear or frame of the machine, especially since this machine is not provided with any springs or other device for taking up the shock of a bad landing.

CURTISS AEROPLANE PARTS—A COMPLETE LIST[1]

1, Engine Section Panel; 2, Wing Panel; 3, Wing Panel, Sparred Beam; 4-5, Aileron, Right & Left; 6, Tail; 7-8, Flipper, Right and Left; 9, Rudder; 10, Front Control, Elevator only; 11, Hydro Front Control, Elevator only; 12-13, Fin, Top & Bottom; 14-15, Non Skid Surface, Headless & Large.

BAMBOOS

16-17, Front, Upper, Right & Left; 18-19, Front, Lower, Right & Left; 20, Front Cross Tie, Headless; 21-22, Front Bamboo Brace, Right & Left; 23-24, Rear, Upper, Right & Left; 25-26, Rear, Lower, Right & Left; 27, Push Rod Bamboo, 45"; 28-29, Bamboo Post, Short & Long.

30, Full Set Rear Bamboos, Wired Complete; 31, Full Tail Equipment, consisting of Rear Bamboos, Posts, Tail, Rudder and Flippers.

POSTS

32, Wing Panel, $\frac{3}{8}''$ x $2\frac{3}{4}''$ x $54\frac{1}{2}''$; 33, Wing Panel, $\frac{3}{8}''$ x $2\frac{3}{4}''$ x $60''$; 34, Engine Section, $1\frac{1}{2}''$ x $2\frac{3}{4}''$ x $54\frac{1}{2}''$; 35, Engine Section, $1\frac{1}{2}''$ x $2\frac{3}{4}''$ x $60''$.

[1] To indicate the exact technical knowledge required in building an aeroplane, a matter quite apart from the obvious dash and daring of the aviator, nothing seems more adequate than to include the list of aeroplane and motor parts.—THE PUBLISHERS.

BIPLANE PARTS

DIAGONAL ASH BRACES, FROM FRONT WHEEL TO ENGINE BED

36-37, Diagonal Ash Brace, Tinned, Right & Left; 38-39, Diagonal Ash Brace, Left & Right; 40-41, Diagonal Ash Brace, Tinned & Ironed, Left & Right.

DIAGONAL SPRUCE BRACE, FROM FRONT WHEEL TO WING PANEL

42-43, Diagonal Spruce Brace, Left & Right; 44-45, Diagonal Spruce Brace, Ironed, Left and Right; 46, Skid; 47-48, Engine Bed, not Tinned, Right & Left; 49-50, Engine Bed, Tinned, Right & Left.

ENGINE BED POSTS. BRACES AND TUBING BRACES ABOVE LOWER PLANE

51-52, Engine Bed Post, Front, Right & Left; 53-54, Engine Bed Post, Rear, Right & Left; 55-56, Engine Bed Brace, Front, Lower, Right & Left; 57-58, Engine Bed Brace, Rear, Lower, Right & Left; 59-60, Engine Bed Brace, Rear, Upper, Right & Left; 61-62, Engine Bed to Surface, Rear, Upper, Right & Left; 63, A Brace to Surface, Front, Upper; 64, Cross Tie Brace under Upper Surface; 65-66, Aileron Brace, Upper, Right & Left; 67-68, Aileron Brace, Lower, Right & Left; 69-70, Seat Post, Right & Left; 71-72, Carburetor Brace, Right & Left.

CHASSIS BRACES. FORKS AND TUBING UNDER LOWER PLANE

73, Cross Tie Rod, Lower, Under Lower Surface; 74, Long Span Brace, Rear Wheel to Rear Wheel; 75-76, Skid Fork, Right & Left; 77-79, Vertical Fork, Front & Rear, Right & Left; 80-81, Leader Fork, Rear, Right & Left; 82-83, M Brace, Right & Left; 84, Y Brace; 85, V Brace, Front, Skid to Diagonal; 86, V Brace Spreader and Bolt, Front; 87, Brace, Center, Skid to Diagonal; 88, V Brace, Center, Skid to Double Seat; 89, V Brace, Rear, Skid to Diagonal; 90-91, Combination Foot Throttle & Brake, Single & Dual.

92, Brake Shoe; 93, Brake Shoe Hinge; 94, Brake Shoe Lug; 95, Brake Shoe Spring; 96, Steering Column, Single; 97, Steering Wheel, Spider, Fork and Bolt; 98, Steering Wheel, Spider, Fork & Column, Assembled & Wired; 99, Steering Column, Dual; 100, Steering Wheel, Spider, Fork & Bolt, Dual; 101, Steering Wheel, Spider, Fork, Bolt & Column, Assembled & Wired, Dual; 102, Foot Rest; 103, Push Rod, Metal, with Swivel End, Dual.

104, Seat, Single; 105, Seat with Fittings for Shoulder Yoke, Single; 106, Seat, Complete with Shoulder Yoke, Whiffle-tree Case and Whiffle-tree, Single; 107, Seat, Double; 108, Seat with Fittings for Shoulder Yoke, Double; 109, Seat, Complete with Shoulder Yoke, Whiffle-tree Cases and Whiffle-tree, Double; 110, Seat, Passenger; 111, Seat Supporting Brace, Passenger; 112, Rear Beam Reinforcing Plates.

113, Cable, $\frac{1}{32}''$; 114, Cable, $\frac{1}{16}''$; 115, Cable, $\frac{3}{32}''$; 116, Cable Casing; 117, Short Circuiting Switch; 118, Snaps, 3''; 119, Main Plane Socket; 120, Main Plane Socket, Wired Complete; 121, Main Plane Plate; 122, Aileron End Wire Connection; 123-124, Aileron Cross Wire Clamp & Clip; 125, Aileron L; 126, Aileron Post Lug; 127, Aileron Brace Wire Connection; 128, Aileron Corner Wire Guide; 129, Aileron Corner Pulley, 3''; 129, Aileron Pulley, 3''.

131, Bamboo Curved Rudder Wire Guide; 132, Skid Safety Wire Connection; 133, Copper Sleeve; 134, Tin Thimbles; 135, Diagonal Ash Brace Iron; 136, Diagonal Spruce Brace Iron; 137-138, Engine Bed Post Plate & Wire Connection; 139, Engine Bed Bolt; 140, Fin L Irons; 141, Fin Hinge; 142-143, Front Control Bracket & L Iron; 144, Hydro Front Control, Brace Lug; 145-146, Hydro Front Control Supporting Post, L & R; 147-148, Hydro Front Control, Supporting Post Lug, Left & Right; 149-150, Hydro Front Control Push Rod & Bracket; 151-152, Hydro Front Control Post & Diagonal Brace; 153, Hydro Splash Boards.

154-155, Flipper Post & Wedge; 156, Flipper Hinge; 157, Flipper Wire Guide, Straight; 158, Rudder Swivel; 159,

BIPLANE PARTS

Curved Corner Wire Guide; 160, Rudder Lever Clip; 161, Rudder Wire Connection; 162, Rudder Wire Guide, Curved; 163-164, Terminals, Short & Long; 165, Turnbuckles; 166, Wheel, 20" x 4", Complete; 167, Wheel, 20" x 4", Less Tire; 168-169, Wheel, 20" x 2½", Complete & Less Tire; 170, Inner Tube, 20" x 4"; 171, Casing, 20" x 4"; 172, Tire, 20" x 2½"; 173, Axle.

174, Gas Tank, to Attach to Engine Bed; 175, Bamboo Brace Clip; 176, Flexible Gasoline Pipe; 177, Radiator; 178, Radiator Brace; 179-180, Propeller, Bolt & Tinned; 181, Propeller, Complete Not Tinned; 182, Cap Screw, Nickel Steel, $5/_{16}$-24 x 1¾; 183, Cap Screw, Nickel Steel, $5/_{16}$-24 x 2¼; 184-185, Spring Washer, ¼ x $3/_{16}$ & $5/_{16}$ x ⅜; 186, Wing Pontoon, Complete; 187, Pontoon Paddles; 188, Hydro Drain Plug; 189, Hydro Braces; 190-191, Hydro Spacing Tube & Bolt, Short & Long.

CHAPTER III

THE CURTISS MOTOR AND FACTORY

THE history of the Curtiss motor goes back to the early days at Hammondsport; it was the keynote of the development of the motorcycle, the airship, the aeroplane, and the hydro. From a crude single-cylinder engine used on an experimental bicycle, the motor has developed to an eight-cylinder engine giving over eighty horsepower, on which the reliability of the Curtiss aeroplane is dependent. Indeed, flight itself, in the history of the world, was delayed until the development of the gas engine made it possible to get a power that was applicable for this purpose, and one that was, at the same time, light enough.

To describe the motor intelligibly to one who has had no acquaintanceship whatever with gas engines would require many chapters, but to those who have ever examined automobile, marine, or other motors, the following technical data will give an idea of the distinctive feature of this aeroplane motor.

MOTOR PARTS

MOTOR DESIGN AND MATERIAL.

Crankshaft:

The crankshaft is supported in five bearings of more than ample size. It is extremely difficult, if not impossible, to design a shaft which will be light enough for aeronautical purposes, and still be sufficiently rigid without a special support. The propeller end of the shaft is supported in two places eleven and three-eighth inches apart, at one end in a plain bearing two and seven-sixteenth inches long and at the other in a combined radial and thrust ball bearing of ample size. This construction is stronger than is the case where the propeller is mounted immediately behind the last main bearing proper or even in some cases carried at a distance of several inches from the bearing without support. Any lack of mechanical or thrust balance is multiplied and transmitted directly to the last crank throw, the tremendous racking and twisting strain thus occasioned causing ultimate failure.

The crankshaft is made of imported Chrome-Nickel steel, properly heat treated. This steel, particularly after heat treatment, has an enormous tensile strength combined with a very high elastic limit and great resistance to fatigue and crystallisation.

Connecting Rods:

The connecting rods are machined from a solid

Chrome-Nickel steel forging, heat treated. The body of the rod is tubular, which cross section gives a maximum strength with minimum weight. Rough forging weighs five pounds; finished weight one pound eight ounces.

Piston:

The piston is long enough to give sufficient bearing surface to sustain the side thrust from the connecting rod and at the same time weighs but two and one-half pounds. The domed head, with properly placed ribs, assures strength. The piston pin bearing is seven-eighth inches diameter by two and three-fourth inches long. Reversing common practice, the pin turns in the piston instead of the rod end, as considerable gain in bearing surface is thus made.

Engineers will appreciate that with a combined piston and rod weight of four and one-half pounds, the strains from twenty-two hundred reversals of motion per minute at normal speed are very slight.

It has three rings together with fourteen oil grooves aiding the rings in retaining compression and assisting the oiling. All pistons are rough turned and then thoroughly annealed before grinding, to insure against warping in service.

The piston rings are of clean springy iron, ground all over. As a ring must be tight on the sides as well as where it comes in contact with the cylinder, there must not be a variation in width of over a quarter thousandth of an inch.

Cylinder:

The cylinder is symmetrical in design, insuring even expansion without distortion.

Valve-in-the-head construction gives an efficient shape of combustion chamber; the compact charge fired in the centre giving quick, complete combustion, and the large valves give free ingress and egress for the gases.

The water jacket is brazed to the cylinder-casting autogenously, the metal being a composition of nickel and copper known as "Monel" metal, which is proof against corrosion.

Cylinders are bored, ground and finished by lapping, to get a glass smooth surface.

Water Circulation:

The water circulation is so carried out that all cylinders are cooled equally, the water pump being divided by a partition which passes water in equal quantities to each set of four, thus avoiding any possibility of a steam-trap on one side causing all the water to pass through the other side. The pump is driven from the crankshaft by a floating joint. The pump shaft is made of a carbon spindle steel.

A portion of the hot water is returned through the carburetor water jacket, which is essential with present day gasoline, particularly in cold weather or high altitudes.

Lubrication:

The lubrication is a combined circulating and splash oiling system. A gear driven oil pump submerged in the oil pan forces a constant stream of filtered oil through the hollow cam shaft bearing, thence to each individual cam shaft bearing, thence to the main crankshaft bearings whence it is forced through the hollow crankshaft and cheeks to the crank pins, the surplus replenishing the oil pan into which the rods dip, thus oiling the cylinder walls by splash and also filling oil pockets on each main bearing, as an additional insurance against their running dry.

The pump is driven off a bevel gear integral with the crankshaft and is of the gear type, being without valves or moving parts other than two simple spur gears. It is entirely enclosed in a fine mesh screen through which the oil must pass to reach the pump.

Valves:

The valves have cast-iron heads reinforced with a perforated steel disc embedded in the cast iron, the whole being electrically welded to a carbon steel stem. The cam shaft is hardened and ground and cams formed integral with the shaft. The cam contour is also ground, the valve timing being exactly the same in each cylinder.

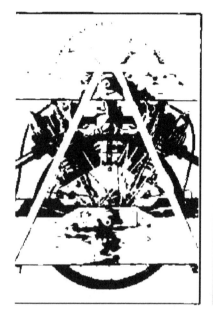

CURTISS MOTORS

(A) The first Curtiss aerial motor; used in Baldwin dirigible. (B) Motor used in both the "White Wing" and "Red Wing." (C) Motor of 1912

AT THE AEROPLANE FACTORY, HAMMONDSPORT
(A) Testing aeroplanes. Gravel on reversed planes tests strength; scale shows wire-strain. (B) Assembly room of factory

MOTOR PARTS

Castings:

The majority of non-moving parts, including the crank case, are cast of special aluminum alloys. Recent laboratory tests have shown tensile strengths of as high as fifty thousand, five hundred pounds per square inch.

Weight:

The weight of model "A" motor alone is two hundred eighty-five pounds—three and eight-tenth pounds per horse-power. The weight of power plant including propeller, radiator, and necessary connections is three hundred forty-seven pounds.

Note that the forty horse-power cylinder motor weighs one hundred seventy-five pounds and gives a thrust of three hundred ten pounds when equipped with a seven foot diameter by six foot pitch propeller turning at nine hundred revolutions per minute. The pitch speed of the propeller at this rate is in excess of a mile a minute.

Gas-Consumption:

The consumption of gas is three-fourths pint per horse-power per hour. The engine can be throttled and consumption reduced in nearly direct ratio to the horse-power developed.

Consumption on full throttle per hour is seven and one-fourth gallons gasoline and one gallon of oil. The oil capacity of the small pan is four gallons; of the large pan, six gallons.

Testing and Power:

Each engine is given an extended run with propeller load. After giving the required standing thrust at the proper speed, the engine is completely torn down for inspection and carbon removed. After assembling, it is given a second test on a water dynamometer, which gives the horse-power developed.

Miscellaneous:

Few people realise that the aeronautical motor is subjected to usage equalled by few internal combustion engines. The average car engine is seldom run on full throttle for extended periods. The marine engine is ordinarily a very heavy, slow speed machine. The aeronautical motor, to run at the high speeds under full load demanded to-day, must of necessity be designed with this fact in mind, and particular attention paid to numerous weaknesses apt to develop under this treatment.

Adding to the above the necessity for minimum weight while still retaining a sufficient factor of safety in all parts, it is evident that an aeronautical motor must be designed as such and not be a modified edition of an automobile engine with a few pounds removed here and there.

MOTOR PARTS

PARTS OF CURTISS MOTOR—A COMPLETE LIST.

1-5, Breather Pipe Cap Screw & Flange, Collar, Cap & Clip; 6, Ball Bearing (Radial); 7-8, Crank Case, Upper Half & Lower Half; 9-10, Crank Case Bolt, Small & Large; 11, Crank Shaft.

12, Cam Shaft; 13-15, Cam Shaft Bearing, Front, Centre, & Rear; 16, Cam Shaft Bearing Sleeve, Rear; 17-18, Cam Shaft Gear & Retaining Screw; 19-20, Cam Shaft Bearing Clamping Screw, Centre, & Retaining Screw; 21, Cam Follower Guide Stud; 22, Cam Follower Guide Screw; 23, Cam Follower; 24-25, Cam Follower Guide & Plug.

26, Cylinder; 27, Cylinder Tie Down Yoke; 28-29, Cylinder Stud, Long & Short; 30, Cylinder Stud Nut; 31-32, Connecting Rod & Bolt; 33, Connecting Rod Bolt Nut; 34, Compression Tee for Oil Pipe; 35, Compression Coupling Sleeve; 36-37, Cable Holder & Screw; 38-39, Cable Tube & End; 40-41, Cable Tube Clip & Screw; 42, Carburetor Water Pipe Clip.

43, Exhaust & Inlet Valve; 44, Exhaust Valve Spring; 45, Felt Oil Retainer for Rear Thrust Bearing; 46, Felt Oil Retainer for Magneto Gear; 47, Gasket for Intake Manifold; 48-49, Gear Case Cover & Screw; 50, Gear Cover Packing Nut; 51, Half Time Gear; 52, Intake Pipe Elbow; 53, Intake Pipe with 2 Union Nuts; 54-56, Intake Pipe Y & Support Base & Cap; 57-62, Intake Manifold, & Bolt,—Bolt Nut,—Cap Screw,—Union Nut,—& Elbow Cap Screw; 63, Intake Valve Spring; 64, Magneto Bracket; 65, Magneto Gear; 66-67, Magneto Bracket Cap Screw, Large & Small; 68, Magneto Base Cap Screw.

69, Main Bearing Stud Nut; 70, Main Bearing Stud, New; 71-73, Main Bearing Cap, Front, Centre & Rear; 74-75, Main Bearing Babbitt, Front, Upper, & Lower; 76-77, Main Bearing Babbitt, Centre, Upper & Lower; 78-79, Main Bearing Babbitt, Rear, Upper, & Lower; 80, Main Bearing Babbitt

Clamping Screw; 81, Main Bearing Liner, Front & Rear; 82, Main Bearing Liner Centre; 83, Main Bearing Liners.

84, Nipple for Oil Pump; 85-86, Oil Pump & Leader Gear Shaft; 87-94, Oil Pump Follower Gear,—Cover,—Drive Pinion,—Screen,—Support Bolt,—Cover Screw,—Follower Gear Bushing,—& Shaft Bushing; 95, Oil Pipe for Pump; 96-97, Oil Pump Compression Coupling & Nut; 98-99, Oil Sight, Base & Glass; 100-101, Oil Sight Glass Guard & Cap; 102, Oil Splash Pan; 103, Oil Bleeder Pipe; 104, Oil Bleeder Pet Cock.

105-107, Piston, Pin & Ring; 108-109, Pump Packing Nut, Large & Small; 110-114, Push Rod, End Bearing Pin Lock Screw,—Spring,—Spring Support,—Forked End,—& End Bearing Pin; 115, Propeller Bolt; 116-121, Rocker Arm,—Support,—Bearing Pin Set Screw,—Tappet Screw,—Support Cap Screw,—& Bearing Pin; 122-124, Spark Plug (Herz)—Gasket,—& Wrench; 125-129, Thrust Bearing, End Clamp,—Lock Ring,—End Clamp Screw,—End Clamp Bolt,—End Thread Bolt Nut; 130, Valve Push Rod; 131, Valve Stem Washer; 132, Valve Stem Lock Washer.

133-135, Water Jacket,—Inlet Nut,—& Inlet; 136, Water Pump; 137-140, Water Pump Shaft,—Support Stud,—Impeller,—& Driver; 141, Water Pump Friction Sleeve; 142-143, Water Pump Friction Washer, Front & Rear; 144-145, Water Pump Bushing, Front & Rear; 146, Water Pump Gasket; 147-149, Water Pump Universal Joint Member, Male,—Female,—& Spring; 150-151, Water Pipe, Right Hand, Bottom,—& Left Hand, Bottom; 152, Water Pipe Outlet Elbow; 153-156, Water Outlet Top Pipes for Cylinders.

A VISIT TO THE FACTORY

A visit to the Curtiss factory is of interest to any one interested in machinery and there you will see the latest machines of all types, from powerful milling machines to a delicate modern "Printograph" that is almost human in its manner of getting out letters and printing, for it is a cross between a printing press and a typewriter. Another unique machine is one that carves out propellers from a laminated block of wood. One arm of this machine runs over a model, and the other, about two feet away, arranged to move exactly with it, and provided with a tool of cutting edge, forms the propeller blade with absolute accuracy, out of a block of wood placed parallel to the model. The cutting tool follows all the complex changes in the surface of the wooden propeller with the greatest ease and rapidity.

The brazing room, where the oxy-hydrogen torch is used to braze metal parts together, and the room where they weld the water jackets on to the cylinders, are places of special interest; the nickel plating room, japanning room, and the room where painting and drying are done, almost complete the tour of the various departments, but there still remain the wood-working shop, boat shop, assembling rooms, where the aeroplanes are put together and completely set up, and the motor testing room, where motors are run for whole days, ten hours at a time, driving an air propeller and showing on scales the amount of thrust given at all times.

Here you may also see a machine to make

"brake tests" of the motors, by which is told how much horse-power the motors give. This machine consists of a large drum with a brake fixed against it and cooled by water so it will not get too hot. This brake absorbs the energy of the motor, which is measured by an arrangement of scales and lever arms.

There is a tremendous racket when the big motors are running at full speed in this small room, and the hillside rings with the roar of their fiery exhaust.

In the laboratory of the factory, where the designs and drawings are made, there is one of the most interesting pieces of apparatus in the whole plant. This is a "wind tunnel," where models of aeroplanes are tested and where experiments are tried to see what occurs in the stream of air. Here tests are made which assist in determining what the best form and shape of objects such as upright posts and exposed parts shall be and where a measure of their relative resistances may be made. The tunnel itself consists of a square box with a propeller or fan mounted at one end to create a draft or current of air which passes through a screen to cause it to assume uniform motion. There is a window in the tunnel through which the observer can see the action of the objects to be tested. Varying the speed of the fan varies the speed of the air current and its pressure, and in this manner the stream-lines of

THE CURTISS FACTORY 307

air under the varying conditions and the effect upon models of different forms and shapes may be studied to enable refinements to be made in the aeroplane's construction.

Down on the shore of Lake Keuka, about a half mile from the factory, are the aeroplane sheds and the flying field. Here is where the aviation school is situated, and where flyers are made. Over the smooth field, the pupils start with the four-cylinder "grass cutters," or machines hobbled so they cannot get but a little way off the ground. They hop, hop, hop, almost all day long, one after the other taking regular turns, and now and again varying the monotony by being called away by the flying instructor to take a real flight in the hydroaeroplane out over the lake to get accustomed to the upper air, and to the high speed of the big machine.

Later in his course of instruction, the student takes out an eight-cylinder machine and flies around in circles over the field until he is able to take the test for his Aero Club of America License, which requires him to make two series of figure eights around two pylons fifteen hundred feet apart, landing each time within one hundred and fifty feet of a mark and rising to an altitude greater than two hundred feet.

This is the goal of the novice, and after his test, the student is ready to fly as far and as fast as he likes. He has become the complete airman.